Organic Reactions and Their Mechanisms

V. K. Ahluwalia

Organic Reactions and Their Mechanisms

Second Edition

**Ane Books
Pvt. Ltd.**

V. K. Ahluwalia
Department of Chemistry
University of Delhi
Delhi, India

ISBN 978-3-031-15697-7   ISBN 978-3-031-15695-3   (eBook)
https://doi.org/10.1007/978-3-031-15695-3

Jointly published with Ane Books Pvt. Ltd.
In addition to this printed edition, there is a local printed edition of this work available via Ane Books in South Asia (India, Pakistan, Sri Lanka, Bangladesh, Nepal and Bhutan) and Africa (all countries in the African subcontinent).
ISBN of the Co-Publisher's edition: 978-93-8826-494-5

This Springer imprint is published by the registered company Springer Nature Switzerland AG
The registered company address is: Gewerbestrasse 11, 6330 Cham, Switzerland

Preface to the Second Edition

The book has been completely revised. The revised edition of the book includes fundamental concepts, equations involved in organic reactions, chemical bonds (ionic and covalent bonds), hybridisation, representation of a chemical reaction and mechanism of organic reactions (including breaking and formation of bonds, reaction intermediates, types of reagents and reactions involved in a chemical reaction). Besides these displacement of bonding electrons involving inductive effect, electromeric effect, mesomeric effect, hyper conjugative effect and resonance have also been incorporated. A number of organic reactions involving formation of intermediates viz. carbocations, carbanions, free radicals, carbenes, nitrenes and benzynes have also been included. Different types of reagents involved in a chemical reaction have also been discussed. Besides these, different types of additional reactions, elimination reactions, substitution reactions, rearrangement reactions, photochemical reactions, oxidation reactions and reduction reactions have also been included. The detailed mechanism of all reactions involved have been discussed.

Finally a large number of questions like multiple choice question, fill in the blanks and short answer questions have been included. All these questions will be useful for a complete understanding of organic reaction mechanisms and also helpful in preparing the candidates for various competitive examinations.

It is hoped that the second edition will be extremely useful to all.

V.K. Ahluwalia

Preface to the First Edition

Organic Reaction Mechanism is written for B.Sc (Gen. & Hons.) students. While Organic Reactions are chemical reactions involving organic compounds, there is no limit to the number of possible organic reactions and their mechanisms? Utmost care has been taken to present the subject matter in a systematic format. Each reaction has a step wise reaction mechanism that explains the causal effects. On reading the text, one would get to understand that Electronic Displacement Effects, Bond Fission, Reaction Intermediate and Different types of Reagents are some of the essentials.

The book has a fine blend of examples and set problems for students and it is hoped that this would facilitate application of knowledge based skills. A core glossary given at the end should be taken as an important study aid.

It is hoped that this book will be extremely helpful to the students in understanding organic reaction mechanism.

The author will appreciate receiving any suggestions and comments for further improvement.

V.K. Ahluwalia

Contents

Organic Reactions and their Mechanisms

1. INTRODUCTION

Unlike inorganic reaction which take place between ions, the organic reactions are basically molecular in nature. An organic reaction takes place by the attack of a reagent on a compound (substrate) to give the products. During a reaction, the bonds in the substrate are cleaved into fragments called reaction intermediates, which are transitory and immediately react with other species (or molecules) present, leading to the formation of a new bond. In fact, an organic reaction involves the cleavage of an existing C—C bond and formation of a new bond to give the products. An organic reaction may be represented as follows:

$$\text{Substrate} \xrightarrow{\text{Reagent}} \left[\begin{array}{c} \text{Intermediate} \\ \text{(transitory)} \end{array} \right] \longrightarrow \text{Products}$$
$$\text{(Reaction intermediate)}$$

The scheme of steps involved in cleavage of a bond and formation of a new bond, leading to the formation of the final products, is referred to as *reaction mechanism*.

Before proceeding with the understanding of the reaction mechanism, it is helpful to understand some of the fundamental concepts.

2. FUNDAMENTAL CONCEPTS

These include the following

- Equation involved in organic reactions
- Chemical Bonds
- Hyberdisation
- Representation of a chemical reaction

© The Author(s) 2023
V. K. Ahluwalia, *Organic Reactions and Their Mechanisms*,
https://doi.org/10.1007/978-3-031-15695-3_1

2.1   EQUATIONS INVOLVED IN ORGANIC REACTION

As a matter of convention, the equations for organic reactions are written with a single reaction arrow ($\rightarrow$) between the starting material and the products formed. The reagent is written on the left side or written on the arrow. The symbols '*hv*' and Δ indicate that the reaction is conducted using light or heat.

or

CCl_4 is the solvent used *hv* shows that the reaction needs light Δ shows that the reaction needs heating.

In case, the reaction is carried out in two steps the steps are usually numbered above or below the arrow. In this case, the first step is carried out before the second step

2.2   CHEMICAL BONDS

Two types of chemical bonds are encountered. These are ionic (or electrovalent) and covalent bond.

2.2.1 Ionic Bonds

These are formed by the transfer of one or more electrons from one atom to another to form ions. An example of the formation of an ionic bond is the reaction of lithium with fluorine.

In the above case, both lithium and fluorine ions attain electronic configuration of nobel gas (helium and neon respectively). The lithium fluoride formed is represented as LiF, this is the simplest formula for ionic compound. Ionic compounds are formed only when atoms of different electronegatives transfer electron to become ions (Electronegatives of Li and F is 1.0 and 4.0 respectively).

2.2.2 Covalent Bonds

Large number of organic compounds are formed due to the ability of carbon to form strong covalent bonds to carbon and other atoms like hydrogen, oxygen, sulphur and nitrogen atoms.

When one or more atoms having same or similar electronegatives react, a complete transfer of electrons do not occur as in the case of ionic bond formation. In such cases, the atoms attain nobel gas configuration by sharing of electrons is called a covalent bond. The formation of a covalent bonds between atoms give products called molecules. The molecules are represented by electron-dot formula or dash formula (the dash represents a pair of electrons shared by the two atoms). Some examples are given below:

$$H\cdot \; + \; H\cdot \; \longrightarrow \; H:H \quad \text{or} \quad H\!-\!H \quad \text{Hydrogen}$$
$$\text{electron} \qquad \text{Dash formula}$$
$$\text{dot formula}$$

$$:\!\ddot{C}l\!\cdot \; + \; \cdot\ddot{C}l\!: \; \longrightarrow \; :\!\ddot{C}l\!:\!\ddot{C}l\!: \quad \text{or} \quad Cl\!-\!Cl \qquad \text{(Chlorine)}$$

$$\cdot\dot{\ddot{C}}\cdot \; + \; 4\cdot H \; \longrightarrow \; H\!:\!\overset{\displaystyle H}{\underset{\displaystyle \ddot{H}}{\ddot{C}}}\!:\!H \quad \text{or} \quad H\!-\!\overset{\displaystyle H}{\underset{\displaystyle H}{\overset{|}{\underset{|}{C}}}}\!-\!H \;\; \text{(Methane)}$$

There is another type of covalent bond which is intermediate between a ionic bond and a pure covalent bond. Such a bond is called **polar covalent bond**. It is formed when two atoms having different electronegatives from a covalent bond, the bonding electrons are shifted to the more electronegative atom.

2.2.3 Hybridisation

Normally, the number of covalent bonds, which an atom forms, is equal to the number of unpaired electrons it has. For example, lithium (electronic configuration $1s^2, 2s^1$) has one unpaired electron ($2s^1$) and so will from one bond. However, beryllium, boron and carbon are exceptions to this generalization. The electronic configuration of these atoms is given below:

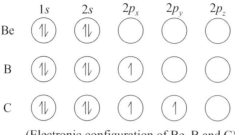

(Electronic configuration of Be, B and C)

On the basis of the above electronic configuration, it is believed that beryllium should be an inert element since it has no unpaired electron. However, beryllium is known to form beryllium fluoride, BeF_2. Similarly, born is expected to form one bond and a carbon atom is expected to form two bonds with other atoms. However, it is known that boron and carbon are known to form compounds like boron trifluoride (BF_3) and methane (CH_4) respectively. Thus, beryllium, born and carbon are *di-*, *tri-* and tetravalent respectively.

In order to explain the 'anomalous' behaviour of these three atoms. It is believed that one of the electrons of 2s orbital (in these atoms) is promoted to the vacant *p* orbital before bond formation takes place (see fig. below). Some energy is required to be supplied to the system to effect this promotion and this energy is more than compensated by the energy released during the formation of covalent bond.

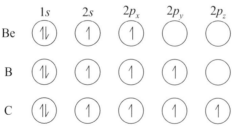

(Promotion of an electron from a 2s atomic
orbital to a 2p orbital in case of Be, B and C)

Thus, the promotion of an electron form a 2s atomic orbital to one of the vacant 2p orbital explains the observed valences 2, 3 and 4 in case of Be, B and C respectively. But the observed geometry of the compounds of these elements is not explained. For example, the orbital of carbon is spherical where as the *p* orbitals are dumb bell shaped and are directed at right angles to each other. In view of this, it is expected that three of the four bonds of carbon would be directed at right angles to each other and the fourth bond formed by the overlap of *s* orbital will not have any definite orientation. It is however, well known that the four bonds formed by carbon are directed towards the four corners of a regular tetrahedron with an angle H—C—H of 109.5°. This problem has been solved by invoking the concept of hybridization.

Hybridization is defined as the redistribution of orbitals of different energy levels in an atom to form a new set of equivalent orbitals. Thus, in case of carbon, promotion of an electron from $2s$ atomic orbital to a $2p$ orbitals will give four equivalent or identical hybridized orbitals. These hybridized orbitals of carbon will be directed towards the four corners of a regular tetrahedron. This type of hybridization of one s and three p orbitals is called sp^3 hybridization. This forms the subject matter of a subsequent section.

In case of carbon, the hybridization is of three types, viz, sp^3, sp^2 and sp hybridization.

2.2.3.1 sp^3 Hybridization

Hybridization (or mixing) of one $2s$ orbital of carbon with three $2p$ orbitals give a set of four equivalent orbitals. As the electrons repel each other, the four orbitals remain at maximum distance from each other and are directed towards the four corners of a regular tetrahedron. This type of hybridization of one s and three p orbitals is called sp^3 hybridization. It is represented as shown below:

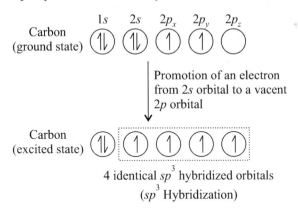

Each of the sp^3 hybridized orbital has 25% s and 75% p character. This is because each sp^3 orbital is obtained from one s and three p orbitals.

2.2.3.2 sp^2 Hybridization

This type of hybridization result by mixing or hybridization of one $2s$ orbital with two p-orbitals. One $2p$ orbital is left unhybridized.

sp^2 Hybridization is explained by taking the example of ethene (CH_2=CH_2). In the case of ethylene, one electron from a $2s$ orbitals gets promoted to a vacant $2p$

orbital ($2p_2$) resulting in the formation of three $2p^2$ hybrid orbitals by hybridization (or mixing) of a $2s$ orbital with $2px$ and $2py$ orbital. This is shown below:

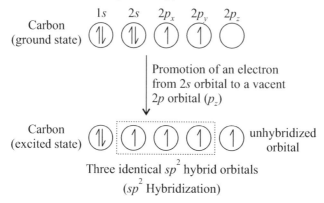

Three identical sp^2 hybrid orbitals
(sp^2 Hybridization)

In ethene, there is a carbon–carbon double bond and each carbon is attached to three other atoms (one carbon and two hydrogen) by the overlap of its three hybrid orbitals. Thus, in ethene, a sigma bond is formed between two carbon atoms by using one sp^2 orbital from each atom. Also one unhybridized p-orbital on each carbon atom overlaps each other sideways producing a π bond. The π electrons are distributed above the plane of the sigma bond. Besides these, two sigma bonds on each carbon atom are formed by the overlap of two sp^2 orbitals in each carbon atom with $1s$ orbital of the hydrogen atom. The three sp^2 hybridized orbitals being in the same plane are directed towards three corners of an equilateral triangle. So this hybridization is called trigonal hybrization, the angle between any two sp^2 orbitals is 120°. Each of the sp^2 hybridized orbital has 33% s and 66% p character. On the basis of the above, ethene molecule is represented as shown below:

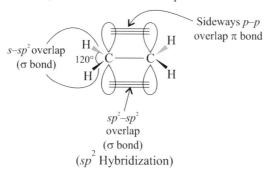

(sp^2 Hybridization)

2.2.3.3 *sp* Hybridization

In sp hybridization, one $2s$ orbital is hybridized (or mixed) with $2p$ orbitals. Two $2p$ orbitals remain unhybridized.

An interesting example of sp hybridization is that of ethyne (HC≡CH). In this case, each carbon of ethyne is bonded with only two other atoms. One electron from a $2s$ orbital gets promoted to a vacant $2p$ orbital ($2p_2$).

Finally, hybridization (or mixing) of one $2s$ and one $2p$ orbital gives two identical sp hybridized orbitals. In this case, two p orbitals ($2py$ and $2pz$) are left umhybridized.

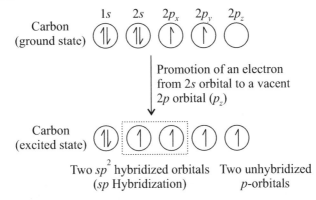

Two sp^2 hybridized orbitals   Two unhybridized
(sp Hybridization) p-orbitals

Ethylene is known to contain a carbon–carbon triple bond. In this case, a sigma bond is formed between two carbon atoms by using one sp orbital from each carbon atom. Two sigma bonds (one on each carbon atom) are formed by the overlap of one sp orbital of each carbon atom with $1s$ orbital of hydrogen atom. Besides these, two unhybridized p-orbital on each carbon atom overlap each other sideways producing two new π bonds, which are distributed above and below the plane of the sigma bond. So the carbon-carbon triple bond in ethyne consists of one sigma bond and two π bonds. Each of the sp bonds has 50% of s character and 50% p character.

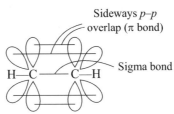

2.3   REPRESENTATION OF A CHEMICAL REACTION

A chemical reaction is represented by different type of arrows depending on the chemical reaction. The most important are the curved arrows since they show which electrons are involved in a reaction. The movement of curved arrows show the forming and breaking of bonds.

- A arrow ($\rightarrow$) is drawn between the starting materials and the products formed in an equation.

- A double reaction arrow ($\rightleftharpoons$) also known as equilibrium arrows are drawn between the starting materials and products in an equilibrium equation.

$$CH_3COCH_2\overset{\overset{\displaystyle O}{\|}}{C}-OEt \rightleftharpoons CH_3\overset{\overset{\displaystyle OH}{|}}{C}=CH\overset{\overset{\displaystyle O}{\|}}{C}-OEt$$

- A double headed arrow ($\leftrightarrow$) is drawn between resonance structures.

- A fully curved arrow ($\curvearrowright$) shows movement of an electron pair.

$$H-A \ + \ :B \rightleftharpoons A: \ + \ H-B^+$$

- A half curved arrow ($\curvearrowright$) also known as fish arrow shows movement of a single electron.

$$A:B \longrightarrow \dot{A} + \dot{B}$$

2.4 MECHANISM OF ORGANIC REACTIONS

The mechanism of organic reactions usually depends on the following factors:
- Electron availability.
- Breaking and formation of bonds.
- Reaction intermediates.
- Type of reagents used in a chemical reaction.
- Type of reactions involved in a chemical reaction.

2.4.1 Electron Availability

In an organic chemical reaction, a substrate molecule is attacked by a reagent, which usually bears a positive or negative charge. In fact, such a reagent will attack the substrate only when it develops an oppositively charge centre. Through a substrate molecule is electrically neutral, yet it must develop some polarity on its carbons for the reaction to take place. This is possible by the displacement of bonding electrons. Such a displacement of electrons in a substrate molecule is due to electronic effect and is referred to as electron displacement effects. Such electronic effects make the availability of electrons in a substrate molecule to make the reaction proceed. There are four type of electronic effects.

- Inductive effect
- Mesomeric effect
- Resonance
- Electromeric effect
- Hyper conjugative effect

2.4.1.1 Inductive Effect

We know that a covalent bond is formed by equal sharing of electrons between the two atoms. If the covalent bond exists between two similar atoms, the electron pair of the bond occupies a central position between the nuclei of the bonded atoms. Such a covalent bond is known as non-polar covalent bond, as in case of hydrogen or chlorine molecules.

$$H : H \qquad\qquad Cl : Cl$$

In case, the covalent bond exists between two dissimilar or different atoms, the electron pair forming σ (sigma) bond is not shared absolutely equally between the two atoms, and gets displaced towards the more electronegative atom of the two. This results in development of a partial negative charge (denoted by $\delta-$) on the more-electro-negative atom and an equivalent partial positive charge ($\delta+$) on the other atom. Thus, in case of C—X bond, where X is more electronegative than C, the state of affairs may be depicted as follows:

$$\overset{\displaystyle}{\underset{\diagup}{\diagdown}}C : X \longrightarrow \overset{\displaystyle}{\underset{\diagup}{\diagdown}}C \overset{}{\rightarrowtail} X \longrightarrow \overset{\displaystyle}{\underset{\diagup}{\diagdown}}\overset{\delta+}{C} - \overset{\delta-}{X} \qquad\qquad ...(1.1)$$

Now consider another bond C—Y where Y is an electropositive atom (i.e., C is more electronegative than Y). In this case, the electron pair forming the C—Y bond is displaced towards the carbon atom. As such, the C atom attains a partial negative charge, $\delta-$ and Y acquires an equivalent partial positive charge, $\delta+$ as shown below:

$$\overset{\displaystyle}{\underset{\diagup}{\diagdown}}C : Y \longrightarrow \overset{\displaystyle}{\underset{\diagup}{\diagdown}}C \overset{}{\leftarrowtail} : Y \longrightarrow \overset{\displaystyle}{\underset{\diagup}{\diagdown}}\overset{\delta-}{C} - \overset{\delta+}{Y} \qquad\qquad ...(1.2)$$

The induction of polarity in a covalent bond as illustrated above is named as *inductive effect*. The polarisation thus induced in the molecule possessing a polarized bond is of permanent nature.

The inductive effect is represented by the symbol $\rightarrowtail$. the arrow head pointing towards the more electronegative atom or group. It is usually indicated as I effect, and is prefixed with a $+$ or $-$ sign, depending upon the direction of electron displacement. Thus, when the substituent linked to the carbon atom is electron attracting (X), it is said to exert a negative inductive effect or $-$ I effect (Eq. 1.1). On the other hand, if the substituent Y bonded to the carbon atom is electron releasing, it exerts a positive inductive effect or $+$ I effect (Eq. 1.2.)

$$\overset{\displaystyle}{\underset{\diagup}{\diagdown}}C \rightarrowtail X \qquad\qquad\qquad \overset{\displaystyle}{\underset{\diagup}{\diagdown}}C \leftarrowtail Y$$
$$\text{–I Effect} \qquad\qquad\qquad\quad \text{+I Effect}$$

Some common substituents/groups causing – I effect (in decreasing order) are noted below:

$$NO_2 > F > COOH > Cl > Br > I > OH > C_6H_5$$

Some common groups causing $+$ I effect (in decreasing order) are noted below:

$$(CH_3)_3C -> (CH_3)_2CH -> CH_3CH_2 -> CH_3 -$$

tert. butyl group   isopropyl group   ethyl group   methyl group

In general, the + I effect of the alkyl groups in the order $3° > 2° > 1°$ groups. In other words, the electron releasing power of a primary carbon ($1°$) is less than that of a secondary carbon ($2°$) which in turn is less than that of a tertiary carbon ($3°$).

The inductive effect gets transmitted along a chain of carbon atoms. Thus, if we consider propyl chloride, the inductive effect makes the α carbon atom acquire a partial positive charge, δ^+, which in turn makes the β carbon atom acquire a still less positive charge than that on α carbon, and so on, the charges on β and γ carbon atoms being represented as $\delta\delta+$ and $\delta\delta\delta+$, respectively.

$$\overset{\delta\delta\delta+}{CH_3}—\overset{\delta\delta+}{CH_2}—\overset{\delta+}{CH_2} \rightarrow \overset{\delta-}{Cl}$$
$$\quad\gamma\qquad\beta\qquad\alpha$$

The inductive effect is very important in organic chemistry in the sense that a number of facts can be explained on the basis of this effect. Some of these are discussed below:

1. **Effect on bond length:** We now know that inductive effect leads to ionic character in the bond. The increase in inductive effect usually decreases the bond length. For example, the bond length between carbon and halogen in alkyl halides decreases with the increase in the inductive effect of the halogen atom.

Increase in inductive effect →

Alkyl halide:	$CH_3 \rightarrow I$	$CH_3 \rightarrow Br$	$CH_3 \rightarrow Cl$	$CH_3 \rightarrow F$
Bond length:	2.14 Å	1.94 Å	1.78 Å	1.38 Å

Decrease in bond length →

2. **Effect of dipole moment:** We now know that inductive effect leads to a dipolar character in a molecule. Therefore, the dipole moment increases with the inductive effect.

Increase in inductive effect →

Alkyl halide:	$CH_3 \rightarrow I$	$CH_3 \rightarrow Br$	$CH_3 \rightarrow Cl$
Bond length:	1.64 D	1.79 D	1.83 D

Increase in dipole moment →

3. **Reactivity of alkyl halides:** The concept of inductive effect clearly reveals that the alkyl halides are more reactive than the alkanes. Further, the order of reactivity of tertiary butyl chloride, isopropyl chloride and methyl chloride on the basis of inductive effect, is as given below:

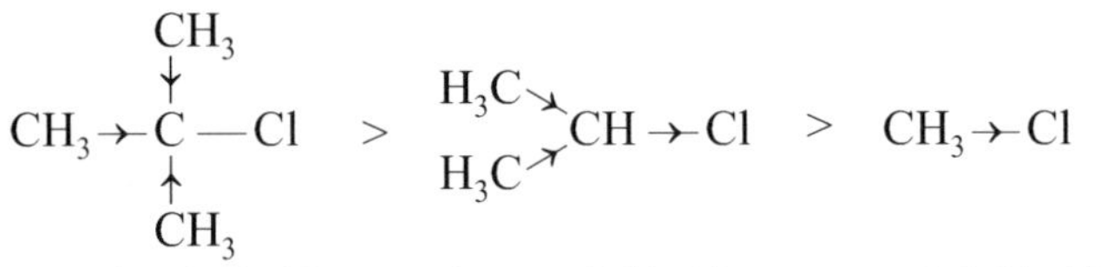

$$CH_3 \rightarrow \overset{\overset{\displaystyle CH_3}{\uparrow}}{\underset{\underset{\displaystyle CH_3}{\uparrow}}{C}} — Cl \quad > \quad \overset{\displaystyle H_3C}{\underset{\displaystyle H_3C}{>}}CH \rightarrow Cl \quad > \quad CH_3 \rightarrow Cl$$

tert. butyl chloride   isopropyl chloride   methyl chloride

As seen above, *tert.* butyl chloride is more reactive (the chlorine atom in *tert.* butyl chloride can be very easily replaced by other atoms) than isopropyl chloride or methyl chloride. This is because of the + I effect of the three methyl groups.

4. **Strength of carboxylic acids:** A carboxylic acid has a tendency to lose a proton. The strength of an acid depends on the ease which it ionises to give a proton. A molecule of carboxylic acid can be represented as a resonance hybrid of the following two structures I and II:

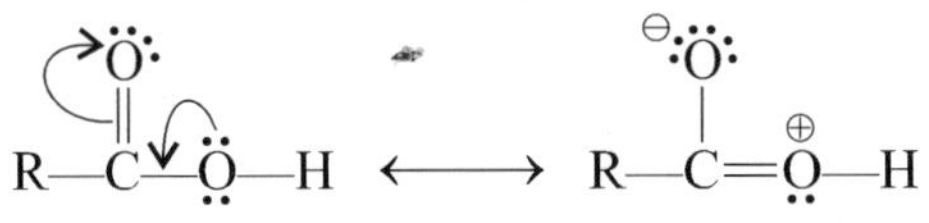

$$R—\overset{\overset{\displaystyle \ddot{O}:}{\|}}{C}—\ddot{O}—H \quad \longleftrightarrow \quad R—C\overset{\ominus\!\!:\ddot{O}:}{=}\overset{\oplus}{\underset{\bullet\bullet}{O}}—H$$

I   II

In structure II, the oxygen atom of the hydroxyl group has a positive charge and so it tends to attract electron pair of the O—H bond (I effect). This results in removal of hydrogen atom as proton and so carboxylic acid behaves as an acid.

$$R—COOH \rightleftharpoons RCOO^- + H^+$$
Carboxylic acid   Carboxylate ion

The resulting carboxylate anion is stabilised by resonance as follows:

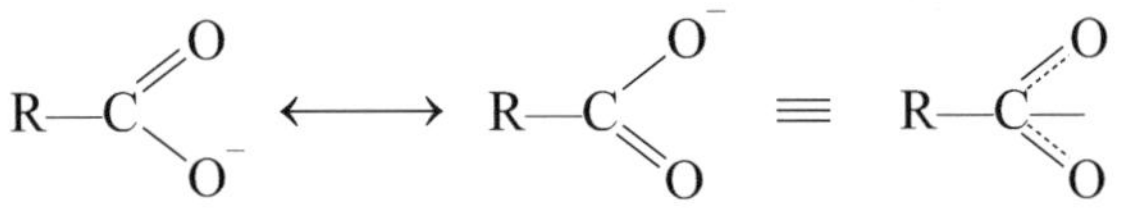

$$R—C\overset{\displaystyle \diagup O}{\underset{\displaystyle \diagdown O^-}{}} \quad \longleftrightarrow \quad R—C\overset{\displaystyle \diagup O^-}{\underset{\displaystyle \diagdown O}{}} \quad \equiv \quad R—C\overset{\displaystyle \diagup O}{\underset{\displaystyle \diagdown O}{}}$$

The existence of carboxylate ion has been confirmed by physical evidences like X-ray and electron diffraction studies.

The strength of the carboxylic acid is based on the intensity of inductive effect and resonance stabilisation of the carboxylate anion. The presence of strongly electronegative group that helps in removal of the hydrogen atom as proton from the hydroxyl group of the acid to form a more easily resonance stabilised carboxylate anion, will make the corresponding carboxylic acid stronger.

The halogenated fatty acids are much stronger than the parent fatty acid (*e.g.,* $ClCH_2COOH$ is more acidic than CH_3COOH). The acidity of the halogenated fatty acids increases with the increase in the electronegativity of the halogen present. Table 1.1 gives the pK_a values of some of the acids. Lower the pK_a value, more stronger will be the acid.

Table 1.1. pK_a values of some carboxylic acids

Carboxylic acid	pK_a value
Propionic acid (CH₃CH₂COOH)	4.76
Acetic acid (CH₃COOH)	4.80
Formic acid (HCOOH)	3.75
Methoxyacetic acid (CH₃OCH₂COOH)	3.53
Iodoacetic acid (ICH₂COOH)	3.16
Bromoacetic acid (BrCH₂COOH)	2.90
Chloroacetic acid (ClCH₂COOH)	2.86
Fluoroacetic acid (FCH₂COOH)	2.57
Cyanoacetic acid (CNCH₂COOH)	2.47

The inductive effect in *di-* and *tri-* halogenated acids is still more marked. Therefore, these acids are more stronger than the corresponding monohalogenated acid. Thus, the acid strength of *tri-, di-* and monochloroacetic acids follows the order:

Trichloroacetic acid ($pK_a = 0.65$) > Dichloroacetic acid ($pK_a = 1.25$) > Monochloroacetic acid ($pK_a = 2.86$)

In fact, cumulative inductive effect of the three chlorine atoms in trichloroacetic acid makes this acid as strong as HCl.

It is found that the inductive effect decreases as the halogen atom (the group responsible for the effect) moves farther from the carboxyl group. Therefore, the strength of the acid decreases as the halogen atom moves from α- to β- to γ- to δ- position in butyric acid. Thus, the acidity of α- β- and γ chlorobutyric acids and *n*-butyric acid follows the order:

Carboxylic acid	pK_a value	
α-Chlorobutyric acid CH₃CH₂CHClCOOH	2.85	
β-Chlorobutyric acid CH₃CHClCH₂COOH	4.05	
γ-Chlorobutyric acid ClCH₂CH₂CH₂COOH	4.50	Decreasing order
n-Butyric acid CH₃CH₂CH₂COOH	4.80	

Similarly, there is a decrease in dissociation constant with an increase in chain length between Cl and COOH group in acids of the type $Cl(CH_2)_n COOH$. Based on this, the acidity decreases in the following order:

Carboxylic acid	pK_a value	
$ClCH_2COOH$	2.85	Decreasing order
$ClCH_2CH_2COOH$	3.98	
$ClCH_2CH_2CH_2COOH$	4.50	

It follows from above that formic acid is a stronger acid than acetic acid. This is because H is less electron repelling than the methyl group. Similarly, propionic acid, butyric acid, isobutyric acid and trimethyl acetic acid are weaker than acetic acid. The acidity of these acids in the following order:

Carboxylic acid	pK_a value	
Formic acid H—COOH	3.75	Decreasing order
Acetic acid $CH_3 \rightarrow COOH$	4.76	
Propionic acid $CH_3 \rightarrow CH_2 \rightarrow COOH$	4.87	
Butyric acid $CH_3 \rightarrow CH_2 \rightarrow CH_2 \rightarrow COOH$	4.80	
Trimethylacetic acid $\begin{array}{c} CH_3 \\ \uparrow \\ H_3C \rightarrow C \rightarrow COOH \\ \downarrow \\ CH_3 \end{array}$	5.05	

The Dicarboxylic acids in general are more acidic than monocarboxylic acids. This is due to the fact that the COOH group has an electron withdrawing inductive effect and the presence of a second COOH group is expected to make it stronger

HCOOH 3.77	HOOC—COOH 1.23
CH_3COOH 4.76	$HOOCCH_2COOH$ 2.83
CH_3CH_2COOH 4.88	$HOOCCH_2CH_2COOH$ 4.19
C_6H_5COOH 4.17	$HOOCC_6H_4COOH$ o-2.98 m-3.46 p-3.51

The above results hold good only if the two carboxyl groups are separated by one or two carbons.

In case of alcohols (unlike carboxylic acids) there is no stabilizing of the alkoxide anion ($RO^{\ominus}$) and so the alcohols are very much less acidic than the carboxylic acids.

The phenols are stronger acids than alcohols (pK_a of phenol is 9.95) but considerably weaker than carboxylic acids.

In case of nitrophenols, the inductive effect falls off with distance ongoing from o- → m- → p-nitrophenol. In addition, there is electron withdrawing mesomeric effect when the nitro group is in o- or p- but not in the m-position. Therefore o- and p- nitrophenols are more acidic than m- nitrophenol. The acidity is further marked by the introduction of more NO_2 groups. Thus 2, 4, 6-trinitrophenol (picric acid) is very strong acid.

	pK$_a$
Phenol	9.95
o-Nitrophenol	7.23
m-Nitrophenol	8.35
p-Nitrophenol	7.14
2, 4-Dinitrophenol	4.01
2, 4, 6-trinitrophenol (Picric acid)	1.02

The effect of introduction of electron donating alkyl group into benzene nucleus is small on the acidity of phenols (cresols)

	pK$_a$
Phenol	9.95
o-cresol	10.28
m-cresol	10.08
p-cresol	10.19

In the case of aromatic carboxylic acids having a double bond are less electron donating compared to the saturated carboxylic acids. This is due to the presence of COOH group being attached to sp^2 hybridized carbon in case of unsaturated aromatic carboxylic acid compared to the attachment of COOH group to sp^3 hybridized carbon in case of saturated aromatic carboxylic acids. On this basis, benzoic acid (pK_a 4.20) is a stronger acid than its saturated analogue, cyclohexane carboxylic acid (pK_a 4.87). As in the case of phenols, introduction of alkyl groups into benzene nucleus has little effect on the strength of benzoic acid

	pK_a
Benzoic acid	4.20
m-methyl benzoic acid	4.24
p-methyl benzoic acid	4.34

On the other hand, the presence of electron withdrawing group (like NO_2 group) into benzene nucleus of aromatic carboxylic acids result in increase of strength. The effect is more pronounced when the electron withdrawing groups are in o- and p- positions (as in the case of phenols).

	pK_a
Benzoic acid	4.20
o-Nitrobenzoic acid	2.17
m-Nitrobenzoic acid	3.45
p-Nitrobenzoic acid	3.43
3, 5-Dinitrobenzoic acid	2.83

The presence of other electron withdrawing groups like Cl, Br, OMe, OH etc also exhibit an electron-donating mesomeric effect, particularly in o- and p- positions. This makes the p- substituted acids weaker than m- substituted carboxylic acids.

The o- substituted carboxylic acids are much more acidic than the m- and p- substituted acids. This is due to the interaction between the adjacent groups. In this case, the intramolecular hydrogen bonding stabilizes the anion by delocalising its charge, which is not present in the case of m- and p- isomers.

o-Hydroxy benzoic acid
(Salicylic acid)   Anion

5. BASIC CHARACTER OF AMINES

In case of amines, the presence of a pair of electrons on Nitrogen (which accepts a proton) makes the amines basic in nature. In fact, the basic strength of amines is dependent on the readiness with which the lone pair of electrons is available for coordination with a proton. Due to the increasing inductive effect of successive alkyl group (which makes nitrogen more negative) it is expected to see an increase in the basic strength on going. $NH_3 \rightarrow RNH_2 \rightarrow R_2NH \rightarrow R_3N$. The relative basic strength is as given below:

$$NH_3 \quad CH_3 \rightarrow NH_2 \quad \begin{matrix} H_3C \\ \diagdown \\ H_3C \diagup \end{matrix} NH \quad \begin{matrix} H_3C \\ H_3C \rightarrow N \\ H_3C \diagup \end{matrix}$$
$$9.25 \qquad 10.64 \qquad\quad 10.77 \qquad\quad 9.80$$

Thus, in case of aliphatic amines, it is found that contrary to the relative basic strength ($NH_3 \longrightarrow RNH_2 \longrightarrow R_2NH \longrightarrow R_3N$), the actual relative basic strength is of the order

$$2° \text{ Amine} > 1° \text{ Amine} > 3° \text{ Amine}$$
$$(CH_3)_2NH \qquad CH_3NH_2 \qquad (CH_3)_3N$$

As seen, introduction of an alkyl group in ammonia (to form 1° amine) increases the basic strength (9.25 → 10.64). However, the introduction of second alkyl group (to from or 2° amine) further increases the basic strength (10.64 → 10.77). In this case the effect of introducing the second alkyl group is very much less. However, introduction of a third alkyl group (to form 3° amine) results in decrease in basic strength (10.77 → 9.80). This is because the basic strength of an amine in water is determined not only on the availability of electrons on N atom, but also by the extent to which the cation formed (by uptake of a proton), can undergo solvation and become stabilized. There is greater possibility of solvation depending on the number of H atoms attached to N in the cation; in these cases the solvation is via hydrogen bonding between these and water.

$$\begin{matrix} H_2O\cdots H \\ | \\ \overset{+}{N}-H\cdots OH_2 \\ | \\ H_2O-H \end{matrix} \quad > \quad \begin{matrix} H_2O\cdots H \\ | \\ R-\overset{+}{N}-R \\ | \\ H_2O\cdots H \end{matrix} \quad > \quad \begin{matrix} H_2O\cdots H \\ | \\ R-\overset{+}{N}-R \\ | \\ R \end{matrix}$$

$$\xrightarrow{\hspace{4cm}}$$
Decrease in stabilization by solvation

Though on going in the series $NH_3 \rightarrow RNH_2 \rightarrow R_2NH \rightarrow R_3N$, the basicity, is increased due to inductive effect, but there is progressively less stabilization of the cation, which is responsible for decrease in basicity. If in case the basicity is determined in nonpolar solvent (like chlorobenzene), there is no possibility of solvation and the basicity of butyl amine is of the order:

$$Bu_3N > Bu_2NH > BuNH_2$$

The presence of electron-withdrawing groups (like Cl, NO_2) close to the basic centre decreases the basicity, which is due to the electron withdrawing inductive effect of the substituents.

In case all the hydrogens in trimethyl amine are replaced by F we get:

$$F_3C-\overset{F_3C}{\underset{F_3C}{\diagdown}}N:$$

This compound is found to be non-basic due to the powerful electron withdrawing effect of three CF_3 groups.

In case of aromatic amines, it is found that aniline is a weaker base (pK_a 4.62) than ammonia (pK_a 9.25) or cyclohexylamine (pK_a 10.68). This is because in aniline, N is bonded to an sp^2 hybridized carbon atom, but more significant is that the lone pair on N (in aniline) interacts with the delocalised π orbitals of the nucleus and so is less available for protonation than in aliphatic amines.

In the anilium cation (obtained by protonation of aniline), the electron pair on N is not available for any interaction.

Anilium cation

Thus, due to the formation of anilium cation, it is not possible for aniline to take up a proton. So aniline is much weaker base (pK_a 4.62) compared to cyclohexylamine (pK_a 10.68).

The basicity of substituted aniline depends on the inductive as well as resonance effects.

The introduction of electron donating group (+ I inductive effect) on N atom of aniline results in small increase in pK_a.

$$\underset{4.62}{C_6H_5NH_2} \qquad \underset{4.84}{C_6H_5NHMe} \qquad \underset{5.15}{C_6H_5NMe_2}$$

However, introduction of more powerful (electron withdrawing) inductive effect like NO_2 has more influence on the basicity. The electron withdrawal is more intensified when the NO_2 group is in o- or p- position. The nitroanilines are found to have the following pK_a values.

$$PhNH_2 \quad O-NO_2C_6H_4NH_2 \quad m-NO_2C_6H_4NH_2 \quad p-NO_2C_6H_4NH_2$$
$$\quad 4.62 \qquad\qquad 0.28 \qquad\qquad\qquad 2.45 \qquad\qquad\qquad 0.98$$

2, 4, 6-Trinitro–N, N–dimethyl aniline is found to be 40,000 times stronger a base than 2, 4, 6-trinitroaniline though there is little difference in case of aniline and N, N-dimethyl aniline. The higher basicity of 2, 4, 6-trinitro-N, N-dimethyl aniline is because the NMe_2 group being sufficiently large interferes sterically with the larger NO_2 group in ortho position. Rotation about ring-carbon to N bonds permit the O atom of NO_2 and Me group of NMe_2 to move out of each others way. Thus the mesomeric shift of the unshared electron pair on NMe_2 group to the O atom of NO_2 group via the p-orbitals of the ring-carbon is inhibited. In fact, the base-weakening of the three NO_2 groups is due to their inductive effects.

2,4,6-Trinitro-N,N-dimethyl aniline

2.4.1.2 Electromeric Effect

When a compound containing multiple bonds (double or triple bond) is exposed to attack by an electrophilic reagent, a pair of electrons is transferred completely from one atom to the other. As a result, the atom to which the electron pair is transferred becomes negatively charged and the other atom becomes positively charged. Thus,

This is purely a temporary effect. As soon as the reagent is removed, the polarised molecule reverts back to its original state.

This effect, which causes temporary polarisation in a substrate molecule containing a multiple bond by shifting of an electron pair from one atom to the other under the influence of an electrophilic reagent, is called *electromeric effect*.

Table (1.2) compares the various features of inductive and electromeric effects.

Table 1.2. Comparison of inductive and electromeric effects.

Inductive effect	Electromeric effect
1. Takes place in molecules containing single bonds.	Takes place in molecule containing multiple bonds ($C=C$ or $C\equiv C$)
2. Takes place under the influence of a substituent attached to the terminal carbon atom.	Takes place only when the molecule is exposed to attack by an electrophilic reagent.
3. Polarity in the molecule is caused by the displacement of bonding electron pair from one atom towards the other. The carbon atom, towards which the bonding electron pair is displaced, develops a partial negative charge ($\delta-$) and the other carbon develops a partial positive charge ($\delta+$) (in case the substituent is electron attracting)	Polarity is caused by complete transfer of an electron pair to one of the two atoms joined by a multiple bond. The carbon atom that gains the electron pair has a positive unit charge (+I) while the other acquires a negative unit charge (–I).
4. It is a permanent effect as it is related to the structure of the substrate molecule.	It is a temporary effect as it disappears on removal of the reagent.

Let us consider the case of a substrate (containing a double bond) exposed to an electrophile. We know that a double bond consists of one σ bond and one π bond (Fig. 1.2), and that the π bonding electrons on a π bond are loosely held and so are easily *polarisable*. In the presence of a charged reagent (electrophilic reagent), the symmetry of the π electron cloud is disturbed, and the electrons of the π bond are completely polarised or transferred to one of the constituent carbon atoms.

Fig 1.2. In the presence of an electrophile, the symmetry of electron cloud is completely disturbed.

Due to the above change, the carbon atom, to which the electron pair is transferred, acquires a negative charge and the other carbon atom gets a positive charge as shown below:

In case of a symmetrical molecule like ethylene, the transfer of electron pair can take place in either direction.

$$H_2C=CH_2 \xrightleftharpoons{\text{Attacking reagent}} H_2\overset{+}{C}-\overset{..}{C}H_2 \quad or \quad H_2\overset{..}{C}-\overset{+}{C}H_2$$

However, in an unsymmetrical molecule, the direction of migration of electron pair depends on the structure of the substrate. Consider, for example, propylene molecule; in this case, the +I effect of the electron repelling methyl group directs the electromeric effect to take place as in (Eq. 1) and not as (Eq. 2), which is opposed by the $+$ I effect of the methyl group.

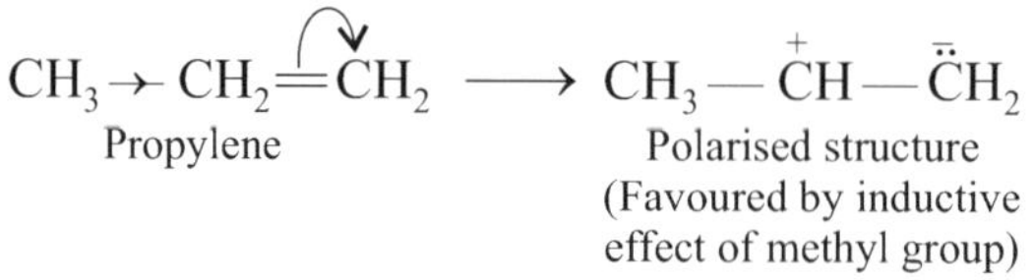

$$CH_3 \rightarrow CH_2 \!\!=\!\! \overset{\frown{\vee}}{CH_2} \longrightarrow CH_3 \!-\! \overset{+}{CH} \!-\! \overset{..}{CH_2} \qquad \qquad ...(1)$$
$$\text{Propylene} \qquad\qquad \text{Polarised structure}$$
$$\text{(Favoured by inductive}$$
$$\text{effect of methyl group)}$$

$$CH_3 \rightarrow \overset{\frown{\vee}}{CH_2} \!\!=\!\! CH_2 \longrightarrow CH_3 \!-\! \overset{..}{CH} \!-\! \overset{+}{CH_2} \qquad \qquad ...(2)$$
$$\text{(\textit{Opposed} by inductive effect of methyl group)}$$

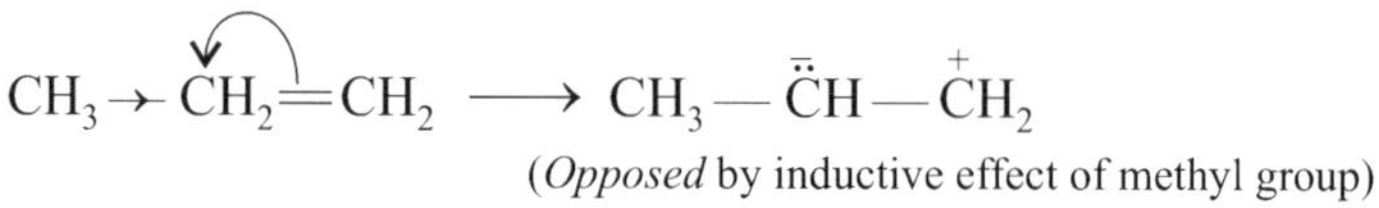

Thus, in the above example of propylene, both inductive effect and electromeric effect support each other. However, there are some cases in which inductive and electromeric effects oppose each other, in such cases, the electromeric effect usually overcomes the inductive effect. For example, in vinyl bromide, the electromeric effect may operate in the following two ways (Eq. 3 and Eq. 4):

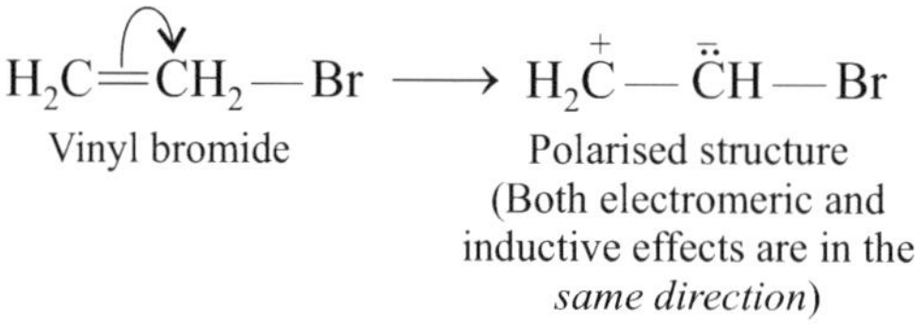

$$H_2C \!\!=\!\! \overset{\frown{\vee}}{CH_2} \!-\! Br \longrightarrow H_2\overset{+}{C} \!-\! \overset{..}{CH} \!-\! Br \qquad \qquad ...(3)$$
$$\text{Vinyl bromide} \qquad\qquad \text{Polarised structure}$$
$$\text{(Both electromeric and}$$
$$\text{inductive effects are in the}$$
$$\textit{same direction})$$

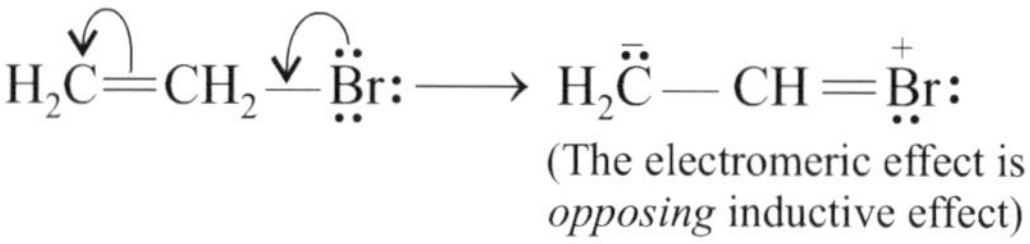

$$H_2\overset{\frown{\vee}}{C} \!\!=\!\! CH_2 \overset{\vee}{-} \overset{..}{\underset{..}{Br}}: \longrightarrow H_2\overset{..}{C} \!-\! CH \!\!=\!\! \overset{+}{\underset{..}{Br}}: \qquad \qquad ...(4)$$
$$\text{(The electromeric effect is}$$
$$\textit{opposing} \text{ inductive effect)}$$

Here, the bromine atom releases an electron pair to be donated to the adjacent carbon atom (as in 4). such and electron transfer from the octet of an atom to that of the another atom without breaking its bond is known as *conjugate effect*. Since the conjugate effect is stronger, it suppresses the inductive effect and so vinyl bromide undergoes electromeric effect as shown in (4).

We have seen above what happens if the multiple bond is present between two similar atoms. However, if the multiple bond is present in two dissimilar or different atoms, the electromeric effect will take place in the direction of the more electronegative atom. For example, in case of carbonyl group, the electromeric effect will take place as follows:

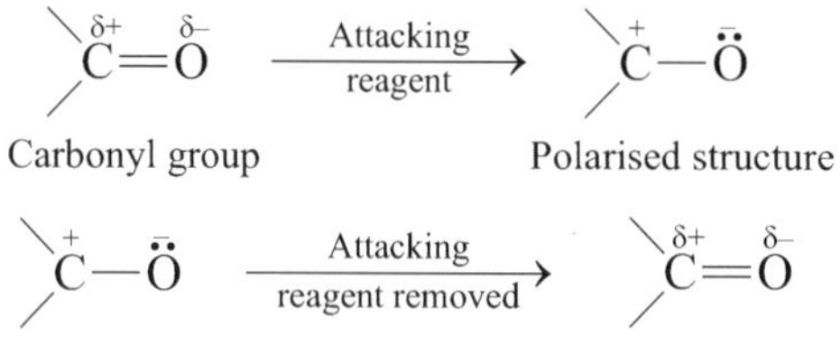

$$\overset{\delta+}{\underset{}{C}} \!\!=\!\! \overset{\delta-}{O} \quad \xrightarrow[\text{reagent}]{\text{Attacking}} \quad \overset{+}{C} \!-\! \overset{..}{\underset{..}{O}}$$
$$\text{Carbonyl group} \qquad\qquad\qquad \text{Polarised structure}$$

$$\overset{+}{C} \!-\! \overset{..}{\underset{..}{O}} \quad \xrightarrow[\text{reagent removed}]{\text{Attacking}} \quad \overset{\delta+}{C} \!\!=\!\! \overset{\delta-}{O}$$

2.4.1.3 Mesomeric Effect

Like inductive effect, the mesomeric effect also causes permanent polarisation. However, unlike inductive effect which operates in molecules having single bond (σ bonds), the mesomeric effect operates only in those systems which have an unsaturated conjugated system. Therefore, this effect is also referred to as *conjugative effect*.

Consider, for example, the case of a carbonly group whose properties are not satisfactorily explained either by formula I or by formula II. The actual structure is considered as a combination of the two structures called a resonance hybrid. In the hybrid structure (shown as III), π electrons are drawn closer towards oxygen than carbon. The effect is denoted by a double headed arrow ($\leftrightarrow$).

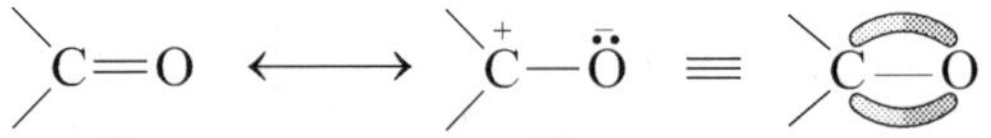

If the carbonyl group is conjugated with an unsaturated system as in case of $CH_2=CH-CH=CH-CH=O$, the π electrons of the C—O bond get displaced towards oxygen (which is mroe electronegative), giving rise to resonance as shown below:

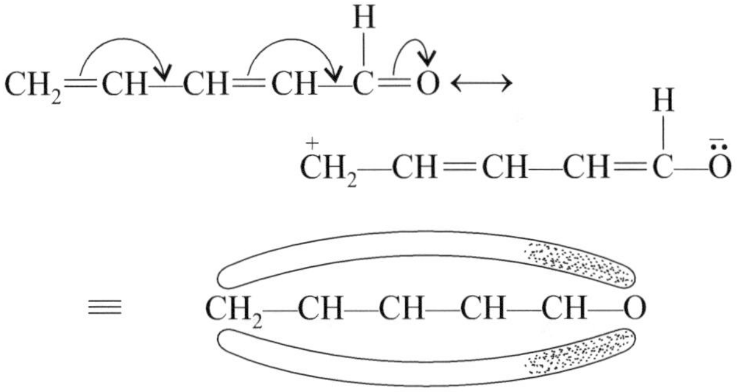

Note that the polarisation is relayed here via the π electrons. The main difference between the mesomeric and the inductive effects is that in the former the terminal carbon is as positive as the first carbon, whereas in inductive effect, the terminal carbon atom is much less positive than the first carbon.

Like the inductive effect, the mesomeric effect (denoted by M) can be + M or – M. Thus, a group or atom has – M effect when the direction of electron displacement is towards it. Some examples of groups showing – M effect are $\overset{\diagdown}{\underset{\diagup}{C}}=O, -NO_2, -C\equiv N, -SO_3H$ etc. The – M effect of nitro group is explained below:

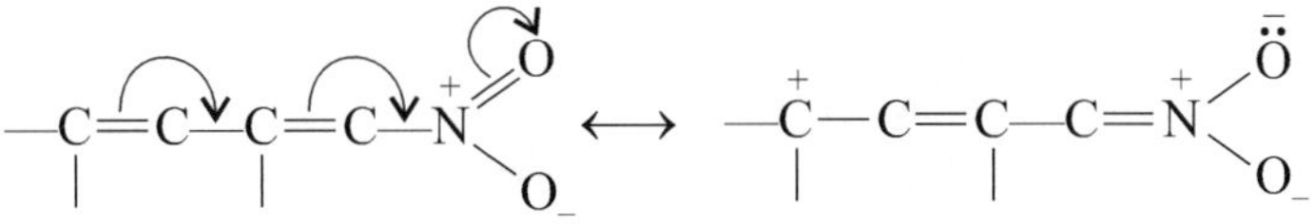

The group or atom shows + M effect if the direction of electron shift is away from it. Some examples of groups showing – M effect are:

$$\overset{\cdot\cdot}{\underset{\cdot\cdot}{O}}-H, \qquad \overset{\cdot\cdot}{\underset{\cdot\cdot}{O}}-R, \qquad \overset{\cdot\cdot}{N}\overset{H}{\underset{H}{\diagup}}, \qquad \overset{\cdot\cdot}{\underset{\cdot\cdot}{S}}-R$$

Note that such groups have a lone pair of electrons and furnish the pair for conjugation with attached unsaturated system. For example, if NH_2 groups is attached to the end carbon of a conjugated system, the electron displacement is represented as:

$$-C=C-C=C-\overset{\cdot\cdot}{N}H_2 \quad \longleftrightarrow \quad -\overset{\cdot\cdot}{C}-C=C-C=\overset{+}{N}H_2$$

The low reactivity of halogen attached to a doubly bonded carbon in vinyl halides and aryl halide can be explained on the basis of + M effect of the halogen atom.

$$H_2C=CH-\overset{\cdot\cdot}{\underset{\cdot\cdot}{Br}}: \quad \longleftrightarrow \quad \overset{\cdot\cdot}{H_2C}-CH=\overset{+}{\underset{\cdot\cdot}{Br}}:$$

(Vinyl bromide)

(Bromobenzene)

Some points of comparison between inductive effect and mesomeric effect are given in Table 1.3.

Table 1.3. Comparison of inductive and mesomeric effects

Inductive effect	Mesomeric effect
1. It operates in molecules containing single bond.	It operates in unsaturated, especially conjugated, compounds which have electron-withdrawing or electron-donating groups attached to the terminal carbon.
2. It involves electrons in the σ bond.	It involves electrons in the π bond.
3. The electron pair is slightly displaced and so, only partial charges (δ+ or δ–) are developed.	The electron pair is transmitted completely and, so full charge (positive and negative) is created.
4. It is transmitted only over a short distance.	It gets transmitted from one end to the other in large molecules having conjugated double bonds.
5. The charge on the terminal carbon atom in a large molecule having 3 – 4 carbon atoms is very much less as compared to the charge on the first carbon atom.	The charge on the terminal carbon atom is of the same intensity as that on the first carbon atom.

2.4.1.4 Hyperconjugative Effect

We know that the inductive effect of alkyl groups is in the order:

Tertiary > Secondary > Primary

However, if the alkyl group is attached to an unsaturated system $\left(\diagup C = C \diagdown \right)$ the above order is reversed. In such systems, the alkyl group becomes capable of electron release in a way which is entirely different from the inductive effect. The mechanism by which electron release take place if the alkyl group is attached to an unsaturated system, is known as *hyperconjugation*. The essential condition for the operation of hyperconjugation is that a carbon–hydrogen (C—H) bond must be present at the α position to the double bond.

For example,

$$-\overset{|}{\underset{|}{C}} - C = C \diagdown \longleftrightarrow -C = C - \ddot{C} \diagdown + H^{+}$$

According to the Baker and Nathan, hyperconjugative effect takes place by the interaction of the electrons of the carbon-hydrogen bond with the π electrons in the double bond. This is best illustrated by the example of propene which has three C—H bonds of the methyl group at α position to the double bond.

$$H - \overset{H}{\underset{H}{C}} - CH = CH_2 \longleftrightarrow H - \overset{H^+}{\underset{H}{C}} = CH - \ddot{C}H_2 \longleftrightarrow$$

$$H^+ \quad \overset{H}{\underset{H}{C}} = CH - \ddot{C}H_2 \longleftrightarrow H - \overset{H}{\underset{H^+}{C}} = CH - \ddot{C}H_2$$

(Hyperconjugative effect in Propene)

On the basis of the above, it is concluded that greater the number of C—H bonds at the α carbon atom to the unsaturated system, the greater would be the electron release towards the terminal carbon atom. On this basis, the hyperconjugative release follows the order:

$$H - \overset{H}{\underset{H}{C}} - \quad > \quad H_3C - \overset{H}{\underset{H}{C}} - \quad > \quad H_3C - \overset{CH_3}{\underset{H}{C}} - \quad > \quad H_3C - \overset{CH_3}{\underset{CH_3}{C}} -$$

Methyl group	Ethyl group	Isopropyl group	Tertiary butyl group
(Primary)		*(Secondary)*	*(Tertiary)*

2.4.1.5 Resonance

The concept of resonance was introduced in connection with the structural elucidation of benzene. It was found that the following two structures can be assigned to benzene:

I   II

As is seen, the positions of the double bonds in both the structures are different. No single structure could explain all the properties of benzene. Both the structures contain 3 single and 3 double bonds. It is known that C=C bond length is 1.33 Å and C—C single bond length is 1.54 Å. However, all C—C bond lengths in benzene are found to be the same (1.39 Å), lying between the bond lengths of single (C—C) bond and double (C=C) bond. Therefore, none of the structures I or II correctly represents benzene. It was, therefore, believed that the true structure of benzene is intermediate between the above two structures; it is type of hybrid that cannot be exactly depicted on paper. Such a molecule is said to be in a state of *resonance*. Thus, resonance is defined as a phenomenon in which two or more structures involving identical positions of atoms can be written for a particular compound. The structure of the actual molecule is expressed as *resonance hybrid*. A double headed arrow (↔) is placed between the various contributing forms. The contributing structures are commonly known as *cannonical structures* or *resonating structures*. Thus, benzene is represented as:

Some of the applications of resonance are as follows

   (*i*) It explains the low reactivity of halogen in vinyl halides and bromobenzene.

  (*ii*) Hyperconjugation can be interpreted in terms of resonance (see the resonating structures of propene)

 (*iii*) It explains the comparative stability of the carbocations

 (*iv*) *Dipole moment of molecules.* In certain cases, the measured dipole moment can only be accounted for in terms of resonance. For example, in vinyl chloride (A), the existence of certain amount of polarity in the bonds and so of the dipole moment can only be explained if structure (B) contributes to the actual structure (resonance hybrid) of the molecule.

$$CH_2 = CH - Cl \longleftrightarrow H_2 \overset{+}{C} - CH = \overset{-}{Cl}$$

            (A)   (B)

(Dipole moment = 1.44 D)

Similarly, resonance also explains the measured dipole moments in cyanides and isocyanides.

$$RC \equiv N \longleftrightarrow R \overset{\delta+}{—} \overset{\delta-}{C} \equiv N$$

$$\overset{+}{R N} \equiv \overset{-}{C} \longleftrightarrow R \overset{..}{N} = C:$$

(v) *Bond length*: Due to resonance, the C—Cl bond length (1.72 Å) is shorter in vinyl chloride than in ethyl chloride (1.78 Å).

(vi) *Strength of acids and bases*: The concept of resonance explains the acidic character of acids and basic character of amines.

(vii) *Stability of free radicals*: The following relative order of stability of free radicals can be explained on the basis of the resonance theory:

tertiary > secondary > primary

Thus, the tertiary free radical (*e.g.*, triphenylmethyl) is the most stable as in this case, the odd electron can delocalise over the central carbon and nine other positions (6 ortho and 3 para) of the 3 phenyl groups).

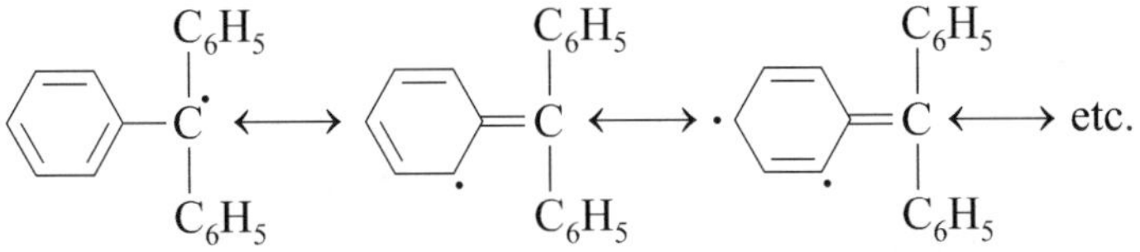

(viii) *Conjugate addition to dienes*: The additions of bromine (1 mole) to a conjugated diene (*e.g.*, butadiene) gives the normal 1, 2-addition product along with the unusual 1, 4-addition product.

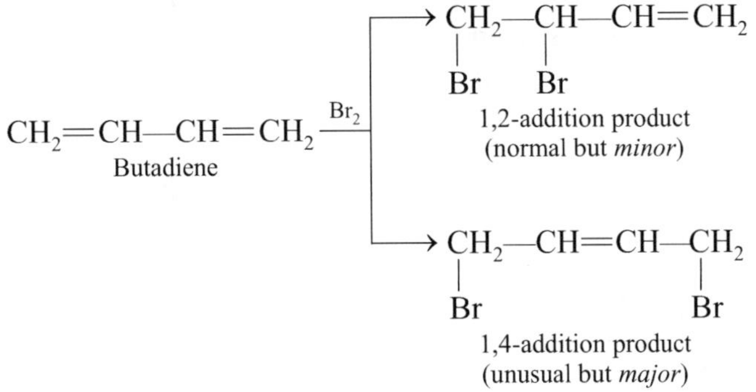

This is explained on the basis of resonance as follows:

The formation of 1, 2-dibromo compound is normal and results from the usual addition to an individual double bond. However, the formation of the unusual 1, 4-addition product (which is obtained in major amount) is believed to arise by the conjugate addition to the diene. It is believed that the bromination takes place in two steps.

In the first step, a carbocation ion is formed in the normal way.

$$CH_2\overset{\oplus}{-}CH-CH=CH_2$$

$$|$$
$$Br$$

2° carbocation

$$CH_2=CH-CH=CH_2 \xrightarrow{Br_2}$$

Butadiene

$$\overset{\oplus}{C}H_2-CH-CH=CH_2$$

$$|$$
$$Br$$

1° carbocation

Now, we know that a 2° carbocation is more stable than the 1° carbocation and, so, the *second step* takes place with the 2° carbocation which, after resonance stabilisation, gives 1, 4-addition product.

Br
|
$$CH-\overset{\oplus}{C}H \overset{\curvearrowright}{-} CH=CH_2 \longleftrightarrow$$

Br
|
$$CH-CH=CH-\overset{\oplus}{C}H_2$$

$$\downarrow Br^-$$

Br   Br
| |
$$CH-CH-CH=CH_2$$

1,2-addition product
(3,4-dibromo-1-butene)

$$\downarrow Br^-$$

Br   Br
| |
$$CH-CH=CH-CH_2$$

1,4-addition product
(1,4-dibromo-2-butene)

Besides resonance, there is another phenomenon in which two isomeric forms having different functional groups are spontaneously interconverted and can exist in a dynamic equilibrium. This phenomenon is called *tautomerism*. All carbonyl compounds, *viz.* aldehydes, ketones and esters, exhibit tautomerism. The best example of tautomerism is that of ethyl acetoacetate.

O   O
|| ||
$$CH_3C-CH_2-C-OC_2H_5 \rightleftharpoons$$

Keto form

OH   O
| ||
$$CH_3-C=CH-C-OC_2H_5$$

Enolic form

(Ethyl acetoacetate)

The presence of both *keto* and *enolic* forms of ethyl acetoacetate has been established by colour and other reactions. In fact, it is also possible to separate the two forms.

The difference between tautomerism and resonance is discussed below:

1. In tautomerism, the two forms of a compound having different structures actually exist and can be isolated as independent molecules. On the other hand, the contributing forms of resonance has no independent existence. Resonance accounts for the stability of the compounds.

2. The structures of tautomers differ mainly in the position of at least one atom relative to the other atoms in the molecule. For example, in the enol

form of ethyl acetoacetate, H is on the oxygen while in the keto form it is on the carbon.

In resonance, the main difference in the canonical forms is in the position of electrons, all atoms in each from have the same relative position.

2.4.2 Breaking and Formation of Bonds

In an organic reaction, the most important is the breaking of the bond (bond fission) and making of a new bond (bond formation).

2.4.2.1 Bond Fission

It is well known that a covalent bond (σ bond) is formed by sharing of a pair of electrons between the two atoms. Bond fission or cleavage of the bond takes place when the two atoms are separated from each other. The fission of a bond can take place in the following two ways:

2.4.2.1.1 Homolytic Bond Fission

In this type of bond fission, the separating atoms may take one electron each. Such a fission is symmetrical or homogenous and is called *homolytic fission* or *homolysis*.

Homolysis is represented as a single electron shift half arrow head (called fish arrow). The two fragments obtained as a result of *homolytic* fission carry an odd electron each and are called *free radicals*.

$$A-B \quad \text{or} \quad A:B \longrightarrow \dot{A} + \dot{B}$$
$$\text{Free radicals}$$

The free radicals are highly reactive intermediates and tend to pair up with another available electron. These radicals are produced by applying some kind of energy (say, heat or light) on the substrate molecule. For example, chlorine molecule is cleaved homolytically in presence of UV light or by heat giving rise to two chlorine radicals which are highly reactive.

$$\underset{\text{Chlorine molecule}}{Cl:Cl} \xrightarrow[\text{or Light}]{\text{Heat}} \underset{\text{Chlorine free radicals}}{\dot{Cl} + \dot{Cl}}$$

As seen, homolysis, involves a single electron shift. However, homolytic fission is also possible by fission of two bonds as illustrated below:

$$\underset{H}{\overset{H}{\underset{X}{C}}}\overset{Y}{\longrightarrow} \underset{H}{\overset{H}{C:}} \quad + \quad \dot{X} + \dot{Y}$$
$$\text{Carbene}$$

In the above case the intermediate obtained is known as carbene.

2.4.2.1.2 Hetrolytic Bond Fission

In this type of bond fission, both the electrons of the bond are taken over by one constituent atom. Such a fission is called *heterolytic fission* or *heterolysis*. This may be represented as follows:

$$A \odot B \longrightarrow A^+ + :B^- \qquad \qquad ...(i)$$

Here, the curved arrow shows the movement of bonding pair of electrons. The atom, to which the electron pair has moved (B in the present case), aquires a negative change while the other atom left short of electrons (*i.e.,* A) gets positive charge. Thus, a heterolytic bond fission give rise to one positive and one negative ion *viz.*,A^+ and :B$^-$ (see equation 1).

Alternatively, the heterolytic bond fission shown above can also take place in the reverse mode as represented below:

$$A \odot B \longrightarrow A:^- + :B^+ \qquad \qquad ...(ii)$$

In this case, the atom A becomes A$^-$ while the atom B becomes :B$^+$ (see equation 2).

Whether heterolytic fission of the bond takes place as in (1) or (2) depends on the electronegative character of the bond atoms. If A is more electronegative than B, the fission will adopt course (2). On the other hand, if A is more electropositive than B, the fission will occur by course (1).

In heterolytic bond fission, the carbon ion carrying a positive charge is called a *carbocation*, and the carbon ion acquiring a negative charge with a pair of electrons is called a *carbanion*.

2.4.2.2 Bond Formation

Bond formation can take place by the reversal of any of the processes as discussed in section 2.5.1.1. It can also take place by the attack of formed intermediates (free radicals, carbanion) on other species.

$$R\bullet + Br\text{-}Br \longrightarrow R\text{—}Br + Br^\bullet$$

$$R^\oplus + H_2O \longrightarrow R\text{—}OH + H^+$$

For more details see section 2.4.3.1.

2.4.3 Reaction Intermediates

Most of the organic reactions take place via the formation of intermediates, which are of transitory existence and take part in the reaction as soon as they are formed. Being very reactive, these cannot be isolated under normal conditions. Their structures are established by indirect means either chemically or spectroscopically. Sometimes the reaction intermediates can be isolated at very low temperature.

Most of the common reaction intermediates that we come across are carbocations, carbanions, free radicals, carbenes, nitrenes and benzynes.

2.4.3.1 Carbocations

We have known (section 2.4.2.1.2) that heterolytic fission of a C—X bond in an organic molecule, if X is more electronegative than the carbon atom, the X takes away the bonding pair of electrons and assumes a negative charge ($:\overline{X}$). In this case, an ion bearing a positive charge is also formed

$$
\underset{\underset{H}{|}}{\overset{\overset{H}{|}}{R-C}} \ddot{:} X \quad \xrightarrow[\text{Fission}]{\text{Heterolytic}} \quad \underset{\underset{H}{|}}{\overset{\overset{H}{|}}{R-C^+}} \;+\; :\overset{-}{X}
$$

The positively species is known as carbocation. These species were earlier called 'carbonium ion'. However they are now referred to as carbocations. In carbocation, the carbon atom is sp^2 hybridized. It uses all its three hybridized orbitals for formation of bonds with other atoms and the remaining p_z orbital is empty and is perpendicular to the plane of the other three bonds. A carbocation is described as planar (trigonal coplanar) with bond angle of 120°.

(Carbocation)

Since carbocation assumes a planar structure, its formation is not possible in compounds which do not permit attainment of a planar geometry as in bridge head compounds.

The presence of electron releasing group such as alkyl group adjacent to the carbon atom bearing positive charge increases the stability of the carbocation. This explains why tertiary carbocation is more stable than a secondary carbocation, which is more stable than the primary carbocation.

| 3° carbocation | 2° carbocation | 1° carbocation | Methyl carbocation |

Besides the presence of electron releasing group (inductive effect), the stability of carbocation is also due to hyperconjugation. Thus the σ electrons to an C—H bond delocalise into the unfilled p-orbital of the positively carbon atom resulting in spreading of charge all over such bonds.

(Hyperconjugative effect increases the stability of carbocation)

It is known that hyperconjugative effects follows the order $3° > 2° > 1°$, which in turn is related to the stability of the carbocation.

The resonance also explains the stability of the carbocations. The more the canonical structures of a carbocation, the more stable it will be. On this basis, benzyl carbocation is more stable than allyl carbocation.

Allyl carbocation:

Benzyl carbocation:

Thus, the order of reactivity of the three carbocations is:

$$C_6H_5\overset{\oplus}{-}CH_2 > CH_2=CH\overset{\oplus}{-}CH_2 > CH_3CH_2\overset{\oplus}{-}CH_2$$

Reactions of Carbocations

As already stated, the carbocations are very reactive and take part in reactions as soon as they are formed. Some of the reactions which carbocation may undergo are discussed below:

1. A carbocation may lose a proton from the α carbon, giving an olefinic compound. For example,

2. A carbocation may combine with a nucleophile to form a bond. Thus, propyl carbocation on reaction with bromide ion forms propyl bromide. In this reaction, the propyl carbocation accepts a pair of electron from the nucleophile (bromide ion).

$$CH_3\text{—}CH_2\text{—}\overset{\oplus}{CH_2} + \overset{..}{Br} \longrightarrow CH_3\text{—}CH_2\text{—}CH_2\text{—}Br$$

Propyl carbocation   Nucleophile   *n*-Propyl bromide
                     (Bromide ion)

Combination with water (a neutral nucleophile) gives a protonated alcohol. For example,

$$
\begin{array}{ccc}
& CH_3 & \\
& | & \\
CH_3\text{—}\overset{\oplus}{C} & +\overset{..}{O}\text{—H} & \longrightarrow \\
& | & \\
& CH_3\quad H &
\end{array}
\qquad
\begin{array}{c}
CH_3 \\
| \\
CH_3\text{—}C\text{—}\overset{\oplus}{O}\text{—}H \\
| \quad | \\
CH_3\; H
\end{array}
$$

Tert. butyl   Water   Protonated
carbocation   (neutral nucleophile)   *tert.* butyl alcohol

The protonated alcohol may subsequently then lose a proton to produce the corresponding alcohol.

$$
\begin{array}{c}
CH_3 \\
| \quad \oplus \\
CH_3\text{—}C\text{—}O\text{—}H \\
| \quad | \\
CH_3\; H
\end{array}
\xrightarrow{-H^+}
\begin{array}{c}
CH_3 \\
| \\
CH_3\text{—}C\text{—}O\text{—}H \\
| \\
CH_3
\end{array}
$$

Protonated   *Tert.* butyl alcohol
tert. butyl alcohol

3. A carbocation may react with an olefin to produce a bigger carbocation. For example,

$$
\begin{array}{ccc}
CH_3 & CH_3 & \\
| & | & \\
C\text{=}CH_2 + & \overset{\oplus}{C}\text{—}CH_3 & \longrightarrow \\
| & | & \\
CH_3 & CH_3 &
\end{array}
\qquad
\begin{array}{c}
CH_3 \qquad CH_3 \\
| \qquad\quad | \\
CH_3\text{—}C\text{—}CH_2\text{—}C\text{—}CH_3 \\
\underset{\oplus}{\;} \qquad\quad | \\
\qquad\qquad CH_3
\end{array}
$$

2-Methyl-   Carbocation   2,4,4-Trimethyl-
propene   2-pentyl cation

4. A carbocation undergoes molecular rearrangement to produce a more stable carbocation. Thus,

(*a*) A primary carbocation on rearrangement gives a secondary carbocation or a tertiary carbocation.

$$CH_3CH_2\overset{\oplus}{CH_2} \qquad \longrightarrow \qquad CH_3\text{—}\overset{\oplus}{CH}\text{—}CH_3$$

primary carbocation   *secondary* carbocation
(propyl carbocation)   (isopropyl carbocation)

$$
\begin{array}{c}
CH_3 \\
| \qquad \oplus \\
CH_3\text{—}CH\text{—}CH_2
\end{array}
\qquad \longrightarrow \qquad
\begin{array}{c}
CH_3 \\
| \\
CH_3\text{—}\underset{\oplus}{C}\text{—}CH_2
\end{array}
$$

primary carbocation   *tertiary* carbocation
(2-methylpropyl   (*tert.* butyl carbocation)
carbocation)

(b) A secondary carbocation on rearrangement may give a tertiary carbocation.

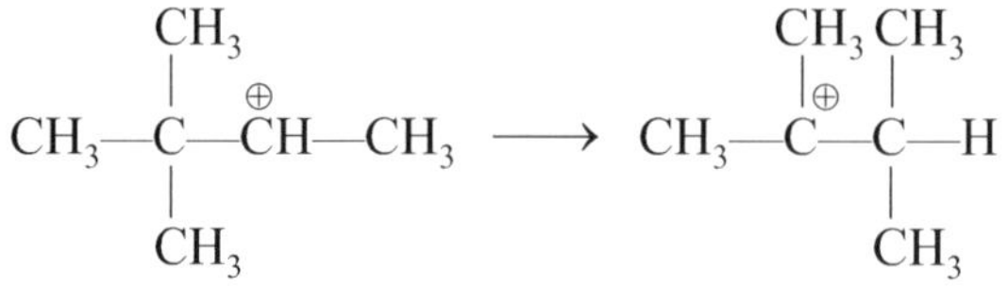

$$\underset{\substack{| \\ CH_3}}{\overset{\substack{CH_3 \\ |}}{CH_3\!-\!C\!-\!\overset{\oplus}{CH}\!-\!CH_3}} \longrightarrow \underset{\substack{| \\ CH_3}}{\overset{\substack{CH_3\,CH_3 \\ | \overset{\oplus}{\ }\,|}}{CH_3\!-\!C\!-\!C\!-\!H}}$$

| 3,3-Dimethyl-2-butyl carbocation | 2,3,3-Trimethyl-2-butyl carbocation |
| (*sec.* carbocation) | (tert. carbocation) |

The above rearrangements take place either by migration of a hydrogen with its pair of bonding electrons, called *hydride shift*, as in case of (a), or by migration of an alkyl group with its pair of bonding electrons, called *alkyl shift*, as in case of (b). A rearrangement of this type, in which the migrating group moves from one atom to the very next atom, is called **1, 2-shift**. The mechanisms for both the cases, *i.e.*, (a) and (b), are outlined as under:

(a) *Hydride shift*:

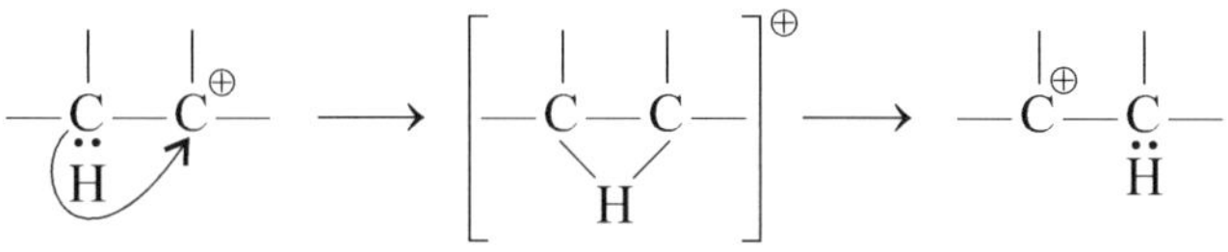

(b) *Alkyl shift*:

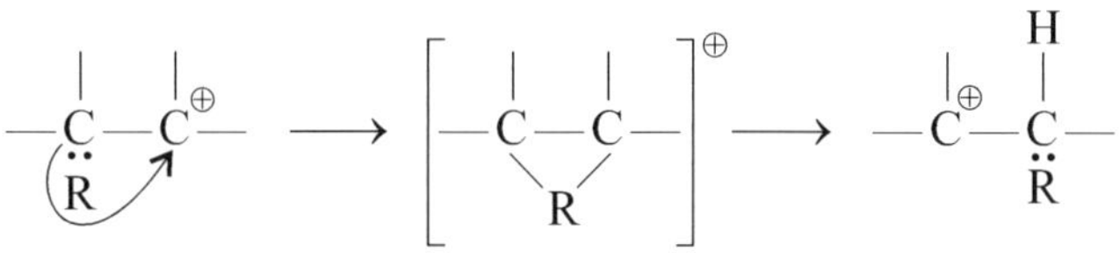

Generation of Carbocations

1. **From alkyl halides:** The alkyl halides, as we already know, on ionisation lead to the formation of carbocation. The solvent plays the role of a reactant and also as solvating medium. The process is catalysed by Ag^+ ions.

$$R\!-\!X \xrightarrow[Ag^+]{Solvent} R^+ + X^-$$

2. **From diazonium cations:** The alkyl diazonium cations are unstable and decompose at room temperature to give carbocation.

$$R\!-\!\overset{\oplus}{N}\!\equiv\!N \longrightarrow R^\oplus + N_2$$

3. **From alcohols:** Alcohols on treatment with conc. acids get protonated and then lose a molecule of water to give carbocation.

$$ROH + H^+ \longrightarrow [R\overset{\oplus}{O}H_2] \longrightarrow R^+ + H_2O$$

This method is now used for the preparation of alkyl benzenes (see alkylation, p. 33).

4. **From acyl halides:** Treatment of acyl halides with anhydrous aluminium chloride generates carbocation.

$$RCOCl \xrightarrow{\text{Anhyd. AlCl}_3} R\overset{\oplus}{C}O + AlCl_4^{\ominus}$$
$$\text{Acylium ion}$$

This, in fact, is Friedel Crafts method for the introduction of —COR group in an aromatic nucleus.

Besides the above method of generation of carbocations, the carbocations have also been postulated as reaction intermediates in a number of reactions like nitrations, halogenation, sulphonation, etc.

Reactions Involving Carbocations

A number of reactions involve the formation of carbocation as intermediates. These include Friedel Crafts-reaction, Baeyer-Villiger oxidation, Pinacol-Pinacolone rearrangement, Wagner Meerwein rearrangement and Beckmann rearrangement.

Friedel-Crafts Reaction

It involves the reaction of aromatic compounds (like benzene) with alkyl halides (the reaction is known as Friedel-Crafts alkylation) or with acyl calorides (the reaction is known as Friedel Crafts acylation) in presence of anhydrous aluminium chloride.

Friedel-Crafts Alkylation

As already stated the reaction of aromatic compounds (like benzene) with alkyl halides in the presence of anhydrous Lewis acid catalyst (like $AlCl_3, FeCl_3,$ $BF_3,$ etc.) results in the formation of alkyl substituted products. This reaction is known as Friedel Crafts Alkylation and is used for the synthesis of alkyl benzenes which are not easily available. Thus, benzene on treatment with ethyl chloride in the presence of anhydrous aluminium chloride gives ethyl benzene. The various steps involved in mechanism of the reaction are as follows:

(a) $CH_3CH_2-Cl + AlCl_3 \longrightarrow \left[CH_3CH_2 \cdots AlCl_4\right] \longrightarrow$

$$CH_3\overset{+}{C}H_2 + AlCl_4^-$$
$$\text{Carbocation}$$

(b)

Benzene   σ-Complex

Ethyl benzene

Friedel crafts reaction is very useful for the synthesis of alkyl substituted benzenes. However, it is difficult to introduce an alkyl group higher than CH_3CH_2 because of skeletal rearrangement taking place within the group. Thus, attempts to prepare *n*-propyl benzene by the alkylation of benzene with *n*-propyl chloride resulted in the formation of isopropyl benzene (cumene).

The formation of isopropyl benzene is explained as follows:

The more stable *secondary carbocation* now takes part in the reaction to give cumene as the product.

In place of alkyl halides, alcohols or alkenes can also be used.

For example,

The carbocations obtained in the above cases are:

$$CH_2CH_2\!\!-\!\!\overset{..}{\underset{..}{O}}H \ + \ H^+ \longrightarrow CH_3\!\!-\!\!\overset{+}{C}H_2 \ + \ H_2O$$
Ethyl carbocation

$$CH_3\!\!-\!\!CH\!\!=\!\!CH_2 \ + \ H^+ \longrightarrow CH_3\!\!-\!\!\overset{+}{C}H\!\!-\!\!CH_3$$
Isopropyl carbocation

Friedel-Crafts Acylation

This reaction involves the introduction of acyl group in an aromatic compound. It is carried out by reacting acyl chloride (*e.g.,* acetyl chloride) with an aromatic compound (like benzene) in the presence of a Lewis acid catalyst (like $AlCl_3$, $ZnCl_2$, $FeCl_3$, BF_3, etc). This reaction is known as *Friedel Crafts Acylation* and is used mostly for the preparation of aryl alkyl ketones. The best example is the preparation of acetophenone by the reaction of acetyl chloride with benzene in the presence of anhydrous aluminium chloride.

$$\bigcirc \ + \ CH_3COCl \ \xrightarrow{\text{Anhyd. } AlCl_3} \ \bigcirc\!\!-\!\!COCH_3$$
Benzene   Acetyl chloride   Acetophenone

The reaction is completed in the following steps:

(a) $CH_3\!\!-\!\!\overset{\overset{\displaystyle Cl}{|}}{C}\!\!=\!\!O \ + \ AlCl_3 \longrightarrow CH_3\!\!-\!\!\overset{+}{C}\!\!=\!\!O \ + \ AlCl_4^-$
Acetyl chloride   Acylium ion

(b) $\bigcirc \ + \ CH_3\!\!-\!\!\overset{+}{C}O \longrightarrow \overset{H}{\underset{}{\bigcirc}}\!\!-\!\!COCH_3$
Benzene   σ-Complex

(c) $\overset{H}{\underset{}{\bigcirc}}\!\!-\!\!COCH_3 \ + \ AlCl_4^- \longrightarrow \bigcirc\!\!-\!\!COCH_3 \ + \ AlCl_3 \ + \ HCl$
σ-Complex   Acetophenone

Friedel Crafts reaction can also be used for the synthesis of aldehydes. Thus, treatment of arenes (benzene) with hydrogen chloride and carbon monoxide in the presence of anhydrous aluminium chloride gives benzaldehyde.

$$\bigcirc \ + \ H\!\!-\!\!\overset{\overset{\displaystyle O}{||}}{C}\!\!-\!\!Cl \ \xrightarrow[\text{Anhyd. } AlCl_3]{(HCl + CO)} \ \bigcirc\!\!-\!\!CHO \ + \ HCl$$
Benzene   Formyl chloride   Benzaldehyde

Baeyer-Villiger Oxidation

The oxidation of ketones to esters with hydrogen peroxide or per acids (RCO_3H) is known as Baeyer-Villiger oxidation. The typical per acids used are peracetic acid, trifluoroperacetic acid, perbenzoic acid, performic acid and *m*-chloroperbenzoic acid (*m*-CPBA). A typical example of Baeyer-Villiger oxidation is the reaction of acetophenone with perbenzoic acid at room temperature.

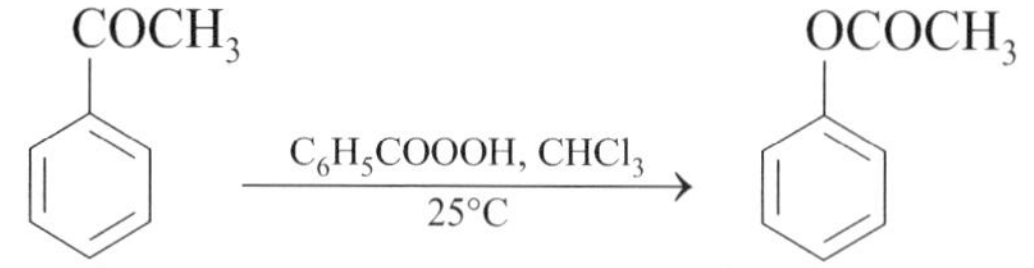

Acetophenone $\xrightarrow[\text{25°C}]{C_6H_5COOOH,\ CHCl_3}$ Phenyl acetate

Mechanism of the reaction:

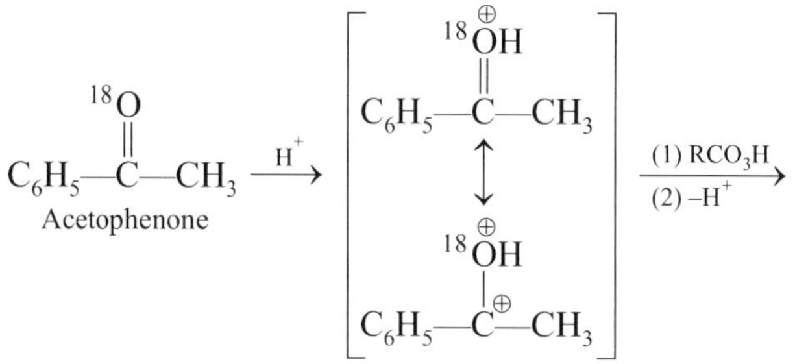

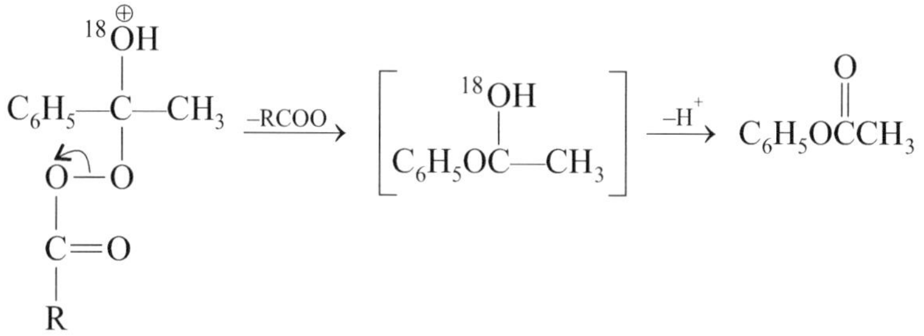

The mechanistic study using ^{18}O labelled ketone has shown that the carbonyl oxygen of the ketone becomes the carbonyl oxygen of the ester and the ester has the same ^{18}O content as the ketone. These observations support the above mechanism for Baeyer-Villiger oxidation.

Pinacol-Pinacolone Rearrangement

The acid catalysed dehydration reaction involving rearrangement is called pinacol-pinacoone rearrangment. For example, 2,3-dimethylbutane-2,3-diol (pinacol) on treatment with sulphuric acid gives 3,3-dimethylbutan-2-one (pinacolone)

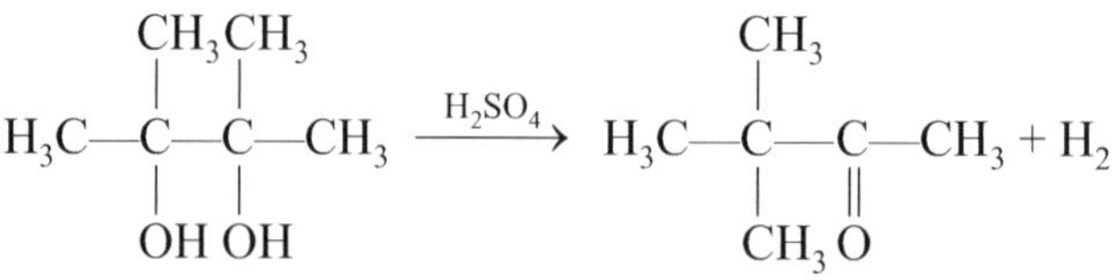

2,3-Dimethylbutane-
2,3-diol (Pinacol)
3,3-Dimethylbutan-
2-one (Pinacolone)

Mechanism of the reaction:

$$
\underset{\text{Pinacol}}{
\begin{array}{c}
\text{CH}_3\text{CH}_3 \\
| \quad | \\
\text{H}_3\text{C}-\text{C}-\text{C}-\text{CH}_3 \\
| \quad | \\
\text{OH OH}
\end{array}}
\xrightarrow{\text{H}^+}
\underset{\text{Oxonium ion}}{
\begin{array}{c}
\text{CH}_3\text{CH}_3 \\
| \quad | \\
\text{H}_3\text{C}-\text{C}-\text{C}-\text{CH}_3 \\
| \quad | \\
\text{OH } \overset{\oplus}{\text{OH}}_2
\end{array}}
\xrightarrow{-\text{H}_2\text{O}}
\underset{\text{Carbocation 1}}{
\begin{array}{c}
\text{CH}_3\text{CH}_3 \\
| \quad | \\
\text{H}_3\text{C}-\text{C}-\overset{\oplus}{\text{C}}-\text{CH}_3 \\
| \\
\text{OH}
\end{array}}
$$

$\downarrow$ 1, 2-shift

$$
\underset{\text{Pinacolone}}{
\begin{array}{c}
\text{CH}_3 \\
| \\
\text{H}_3\text{C}-\text{C}-\text{C}-\text{CH}_3 \\
\| \quad | \\
\text{O} \quad \text{CH}_3
\end{array}}
\xleftarrow{-\text{H}^+}
\underset{\text{Oxonium ion}}{
\begin{array}{c}
\text{CH}_3 \\
| \\
\text{H}_3\text{C}-\text{C}-\text{C}-\text{CH}_3 \\
\| \quad | \\
\text{H}-\overset{\oplus}{\text{O}} \quad \text{CH}_3
\end{array}}
\longleftarrow
\underset{\text{Carbocation 2}}{
\begin{array}{c}
\text{CH}_3 \\
| \\
\text{H}_3\text{C}-\overset{\oplus}{\text{C}}-\text{C}-\text{CH}_3 \\
| \\
:\!\overset{..}{\text{O}}\text{H CH}_3
\end{array}}
$$

Wagner-Meerwein Rearrangement

The rearrangement which occur during reactions involving change in the carbon skeleton via the rearrangement of carbocation intermediates, are collectively known as Wagner-Meerwein Rearrangements. For example, the reaction of HBr on neopentyl alcohol gives *tert.* pentyl bromine (2-bromo-2-methylbutane).

$$
\underset{\text{Neopentyl alcohol}}{
\begin{array}{c}
\text{CH}_3 \\
| \\
\text{H}_3\text{C}-\text{C}-\text{CH}_2-\text{OH} \\
| \\
\text{CH}_3
\end{array}}
\xrightarrow{\text{HBr}}
\underset{\text{2-Bromo-2-methylbutane}}{
\begin{array}{c}
\text{CH}_3 \\
| \\
\text{H}_3\text{C}-\text{C}-\text{CH}_2-\text{CH}_3 \\
| \\
\text{Br}
\end{array}}
$$

Mechanism of the reaction:

$$
\underset{\text{Neopentyl alcohol}}{
\begin{array}{c}
\text{CH}_3 \\
| \\
\text{H}_3\text{C}-\text{C}-\text{CH}_2-\text{OH} \\
| \\
\text{CH}_3
\end{array}}
\xrightarrow[(-\text{H}_2\text{O})]{\text{H}^+}
\underset{\substack{\text{Neopentyl carbocation} \\ (1° \text{ carbocation})}}{
\begin{array}{c}
\text{CH}_3 \\
| \\
\text{H}_3\text{C}-\text{C}-\overset{\oplus}{\text{CH}}_2 \\
| \\
\text{CH}_3
\end{array}}
$$

$\downarrow$ 1,2-methyl shift

$$
\underset{\text{2-Bromo-2-methylbutane}}{
\begin{array}{c}
\text{CH}_3 \\
| \\
\text{H}_3\text{C}-\text{C}-\text{CH}_2-\text{CH}_3 \\
| \\
\text{Br}
\end{array}}
\xleftarrow{\text{Br}^-}
\underset{3° \text{ carbocation}}{
\begin{array}{c}
\text{CH}_3 \\
| \\
\text{H}_3\text{C}-\overset{\oplus}{\text{C}}-\text{CH}_2-\text{CH}_3
\end{array}}
$$

Another example of Wagner-Meerwein rearrangement is the reaction of 3, 3-dimethyl-1-butene with HCl to give 2-chloro-2, 3-dimethylbutane as the major product (60–75%) along with the expected 2-chloro-3,3-dimethylbutane as the minor product (25–40%).

$$CH_3 \quad CH_3$$
$$\longrightarrow H_3C-\overset{\underset{\displaystyle |}{Cl}}{C}-CH-CH_2$$
2-Chloro-2,3-dimethylbutane
(60–75%) (with rearrangement)

$$\overset{CH_3}{\underset{CH_3}{H_3C-\overset{|}{\underset{|}{C}}-CH=CH_2}} \xrightarrow{HCl}$$
3,3-Dimethyl-1-butene

$$\overset{+}{CH_3} \quad Cl$$
$$\longrightarrow H_3C-\overset{\underset{\displaystyle |}{CH_3}}{C}-CH-CH_2$$
2-Chloro-3,3-dimethylbutane
(25–40%) (no rearrangement)

Beckmann rearrangement: A ketoxime on treatment with an acid catalyst (like H_2SO_4, polyphosphoric acid, $SOCl_2$, PCl_5, $C_6H_5SO_2Cl$, etc.) rearranges into a substituted amide.

$$\underset{\underset{\displaystyle Ketoxime}{N-OH}}{R-\overset{\displaystyle ||}{C}-R'} \xrightarrow{PCl_5} \underset{N-Cl}{R-C-R'} \xrightarrow{-\overset{..}{C}l} \underset{R-N}{\overset{+}{C}-R'} \xrightarrow{H_2O}$$

$$\underset{R-N}{\overset{+}{H_2O}-\overset{\displaystyle ||}{C}-R'} \xrightarrow{-H^+} \underset{R-N}{H-O-C-R'} \rightleftharpoons \underset{R-N-H}{O=C-R'}$$
Substituted amide

In Beckmann rearrangement, the group R, which is *trans* with respect to the hydroxyl group, gets migrated and the R radical or groups is never detached completely from the substrate molecule.

2.4.3.2 Carbanions

We have seen that in the heterolytic fission of the C—X bond in an organic molecule, in which carbon is more electronegative than atom X, the carbon takes away the bonding electron pair and acquires a negative charge and X is left with a positive charge on it.

$$\underset{H}{\overset{H}{R-\overset{|}{\underset{|}{C}}-X}} \longrightarrow \underset{H}{\overset{H}{R-\overset{|}{\underset{|}{C}}:}} + \overset{\oplus}{X}$$

(C is more electronegative than X)   Carbanion

The organic ion with a negative charge on the central carbon and having a pair of electrons is called a *carbanion*.

An organic compound having a C—H bond function as an acid (in the classical sense) by donating a proton on treatment with a base forming a carbanion.

$$R_3C—H + B: \longrightarrow R_3C^- : \ + BH$$
$$\text{Carbanion}$$

The carbanion possess an unshared pair of electrons and so is considered as a base. The carbanion can accept a proton to give its conjugate acid. In fact, the stability of the carbanion depends on the strength of the conjugate acid. The weaker the acid, more is its basic strength and less will be the stability of the carbanion.

Table below gives the base (carbanion) obtained from the corresponding acid along with its pK_a value.

Table. Carbanions obtained from the corresponding acid along with their pK_a value

Acid (pK$_a$)	Base (carbanion)
CH_4 (43)	$\overline{C}H_3$
$CH_2=CH_2$ (37)	$CH_2=\overline{C}H$
C_6H_6 (37)	$C_6H_5^-$
$C_6H_8CH_3$ (37)	$C_6H_5CH_2^-$
CF_3H (28)	$\overline{C}F_3$
$HC{\equiv}CH$ (25)	$HC{\equiv}\overline{C}$
CH_3CN (25)	$\overline{C}H_2CN$
CH_3COCH_3 (20)	$CH_3CO\,\overline{C}H_2$
$C_6H_5COCH_3$ (19)	$C_6H_5CO\,\overline{C}H_2$
$CH_2(CO_2Et)_2$ (13.3)	$\overline{C}H(CO_2Et)_2$
CH_3NO_2 (10.2)	$\overline{C}H_2\,NO_2$

The stability of carbanions depends on the following features
- Increase in the S character in the carbanion carbon
- Electron withdrawing inductive effect
- Conjugation of the lone pair of the carbanion with multiple bonds
- Aromatisation

The understanding of the above features is clear from the following discussions:

(a) *Increase in the S-character in the carbanion carbon*

There is increase in the acidity of the hydrogen atom in the sequence $CH_3CH_3 < CH_2\text{—}CH_2 < HC\equiv CH$. The increase in acidity is marked (see table above) as one goes from alkane to alkyne. This reflects the increasing character of the hybrid orbital involved in the sigma bond to hydrogen, *i.e.*, $sp^3 < sp^2 < sp^1$ (25, 33 and 50% S character respectively). The H atom is lost more easily from alkyne followed by alkene and alkane. This in turn result in the stabilization of the resultant carbanion which is of the order.

$$HC\equiv C^- > CH_2\text{=}\overline{C}H > H_3C\text{—}\overline{C}H_2$$

(b) *Electron withdrawing inductive effect.*

The presence of an electron releasing group (like alkyl group) at the end of the chain decreases the stability of the carbanion. On the other hand, the presence of an electron attracting group (like CN, $\diagdown C=O$ etc.) at the end of the chain decreases the stability of the carbanion. Thus,

$$NC\text{—}\!\!\!\leftarrow\overset{\displaystyle |}{\underset{\displaystyle |}{C}}\!\!:^{\ominus} \text{ is } \textit{more stable than } H\text{—}\!\!\!\overset{\displaystyle |}{\underset{\displaystyle |}{C}}\!\!:^{\ominus}, \text{ and}$$

$$H_3C\text{—}\!\!\!\leftarrow\overset{\displaystyle |}{\underset{\displaystyle |}{C}}\!\!:^{\ominus} \text{ is } \textit{less stable than } H\text{—}\!\!\!\overset{\displaystyle |}{\underset{\displaystyle |}{C}}\!\!:^{\ominus}.$$

The electron withdrawing inductive effect is an important factor in determining the stability of the carbanion is well understood by considering the case of HCF_3 ($pK_a = 28$) and $HC(CF_3)_3$ ($pK_a = 11$). In both the cases, the strong electron withdrawing inductive effect of fluorine atoms make the H atoms more acidic and stabilizes the resulting carbanions ($\overline{C}F_3$) and $\overline{C}(CF_3)_3$ by electron withdrawal, since $\overline{C}(CF_3)_3$ has nine fluorine atoms compared to only three in HCF_3, the former viz. $\overline{C}(CF_3)_3$ is more stabilized than $\overline{C}F_3$. It should be noted that in the formation of $\overline{C}Cl_3$ from $HCCl_3$, similar electron-withdrawing inductive effect is operative. However, the electron-withdrawing inductive effect of Cl is less than that of F, $\overline{C}F_3$ is more stabilized than $\overline{C}Cl_3$.

Due to deshielding effect of the electron donating inductive effect of alkyl groups, the stability of the carbanion is of the order:

$$\overline{C}H_3 > RCH_2^- > R_2\overline{C}H > R_3\overline{C}$$

It should, however be noted that the stability sequence for carbanions is exactly the reverse of the stability sequence for carbanions (see section 2.4.3.1).

(c) *Conjugation of the lone pair on the carbanion with multiple bonds.*

The stability of the carbanion depends on the conjugation of the lone pair of carbanion with a polarized multiple bond as seen in the following examples given below:

$$B: \quad \overset{H}{\underset{|}{CH_2}} \rightarrow \overset{\delta+}{C} \equiv \overset{\delta-}{CN} \rightleftharpoons \left[\overset{-}{CH_2} - C \equiv N \longleftrightarrow CH_2 = C = \overset{-}{N} \right] + BH^+ \quad pK_a = 25$$
Acetonitrile

$$B: \quad \overset{H}{\underset{|}{CH_2}} \rightarrow \overset{CH_3}{\underset{|}{C}} = O \rightleftharpoons \left[\overset{CH_3}{\underset{|}{\overset{-}{C}H_2 - C}} = O \longleftrightarrow \overset{CH_3}{\underset{|}{CH_2 = C - \overset{-}{O}}} \right] + BH^+ \quad pK_a = 20$$
Acetone

$$B: \quad \overset{H}{\underset{|}{CH_2}} \rightarrow \overset{+}{\underset{|}{N}} = O \rightleftharpoons \left[\overset{-}{CH_2} - \overset{+}{\underset{|}{N}} = O \longleftrightarrow CH_2 = \overset{+}{\underset{|}{N}} - \overset{-}{O} \right] + BH^+ \quad pK_a = 10.2$$
Nitromethane

Though in each of the above examples, the electron withdrawing inductive effect increases the acidity of the H atoms on the formed carbanion by the delocalisation is of greater significance. Thus, as expected, NO_2 is much more powerful.

(d) *Aromatisation.*

Carbanions are stabilised by aromatisation. As an example, cyclopentadiene (pK_a 16 compared to about 37 for simple alkene) on treatment with a base gives cyclopentadiene anion (a 6π electron system, a $4n+2$ Hückel system, where $n = 1$). The six π electrons in the anion are accommodated in three stabilised π molecular orbitals (like benzene). The cyclopentadiene anion is stabilised by aromatisation and show quasic aromatic stabilisation.

Br: Cyclopentadiene   Cyclopentadiene anion

In similar way, cyclooctatetrane on treatment with potassium gives an isolable, crystalline salt of cyclooctatetraenyl dianion which is also a Hückel $4n + 2\pi$ electron system (n being 2) and shows quasi-aromatic stability. In this case also stabilisation is by aromatisation.

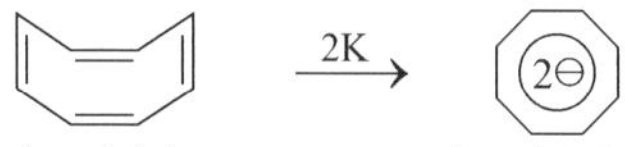

cyclooctatetrane   cyclooctatetraenyl
dianion

(Cyclooctatetraenyl dianion)

Like carbocations, the carbanions are also stabilised by resonance. Thus, benzyl carbanion C_6H_5—$\overset{\ominus}{\ddot{C}H_2}$ is more stable than ethyl carbanion CH_3—$\overset{\ominus}{\ddot{C}H_2}$. The stabilisation by resonance is due to the delcoalisation of the negative charge, which is then distributed over other carbon atoms. The possible canonical forms of benzyl carbanion are given below:

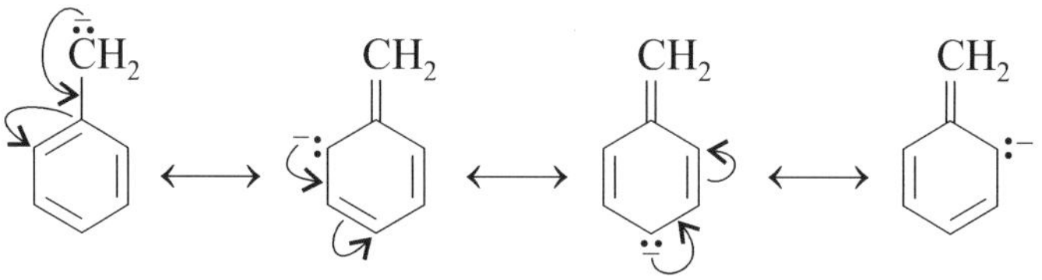

(Four canonical forms)

The stability of various carbanions follows the order as given below:

Benzyl > Vinyl > Phenyl > Cyclopropyl > Ethyl > n-Propyl > Isobutyl > Neopentyl > Cyclobutyl > Cyclopentyl. The stablility of the carbanions carrying a functional group (Z) at α position follows the order:

NO_2 > RCO > COOR > SO_2 > CN ≈ $CONH_2$ > Halogens > H > R

Structure of Carbanions

A simple carbanion of the type R_3C^- can assume a pyramidal (sp^3) or a planar (sp^2) configuration depending on the nature of R.

On the basis of energy grounds, it is believed that the pyramidal configuration is preferred since the unshared electron pair can be accommodated in an sp^3 orbital.

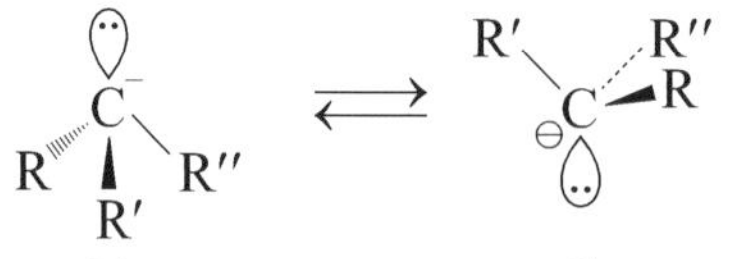

(a)   (b)

The pyramidal configuration (a) is similar to the one adopted by tertiary amines R_3N: with which simple carbanions, $R_3C^\ominus$ are isoelectrone. In carbanions ready inversion of configuration occurs (a ⇌ b) as in the case of amines.

The pyramidal configuration finds support by the observation that the reactions involving the formation of carbanions intermediates at the bridgehead position take place quite readily.

Generation of Carbanions

Following methods are used for the generation of carbanions:

1. **Form alkyl halides.** Reaction of an alkyl halides with magnesium turning in the presence of anhydrous ether as solvent generates carbanion in the form of Grignard reagent.

$$R—X + Mg \xrightarrow{\text{Ether}} R^-MgX^+$$

Grignard reagent behaves like a nucleophile.

2. **Abstraction of H by a base.** This is a very commonly used method for the generation of a carbanion. Thus *p*-nitrotoluene on treatment with a base gives the carbanion.

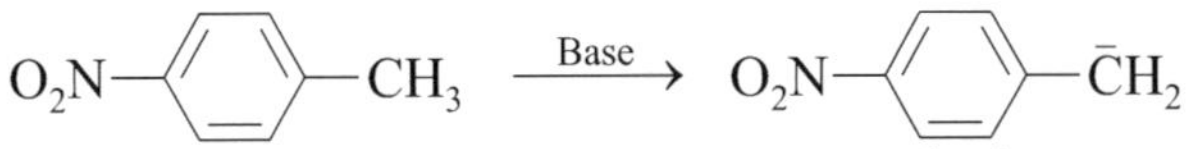

p-Nitrotoluene   Carbanion

The commonly used bases are NaH/ether, aqueous NaOH, K^+t-BuO$^-$/ t-BuOH, N-methylpyrrolidone in ether, etc.

3. **From unsaturated compounds.** Addition of a nucleophile to an unsaturated compound generates a carbanion

Carbanion

Besides the above methods, the carbanions are obtained as intermediates in Aldol condensation, Perkin reaction, Claisen condensation, etc.

Reactions of Carbanions

Carbanions undergo a number of reactions like addition reactions, elimination reactions, displacement reactions, rearrangement reactions, decarboxylations and oxidation.

Addition Reactions

Carbanions, being electrons rich behave like nucleophiles and add to the carbonyl group of aldehydes or ketones

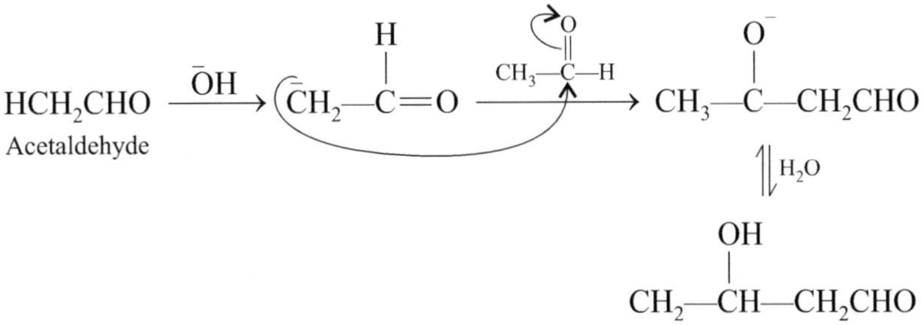

HCH$_2$CHO
Acetaldehyde

This is the well known **aldol condensation.**

Aldehyde or Ketone	Organo metallic compound

Elimination Reactions

Carbanions undergo elimination reactions. As an example, β-phenyl ethyl bromide on treatment with a base gives styrene via the formation of carbanion followed by elimination.

Displacement Reactions

An example of a displacement reaction is the formation of monoalkyl derivative of diethyl malonate or acetylacetone via the intermediate formation of carbanion.

$$CH_2(COOC_2H_5)_2 \xrightarrow{\bar{O}Et} \bar{C}H(COOC_2H_5)_2 \xrightarrow{R-X} RCH(COOEt)_2$$

Diethyl malonate   Carbanion   Monoalkyl derivative

$$CH_3COCH_2COCH_3 \xrightarrow{\bar{O}Et} CH_3CO\bar{C}HCOCH_3 \xrightarrow{R-X} CH_3CO\overset{R}{\underset{|}{C}}HCOCH_3$$

Acetylacetone   Carbanion   Monoalkyl derivative

(Displacement reactions of carbanions)

In these reactions, the formed carbanions act as nucleophiles

Rearrangement Reaction

Reactions involving rearrangement of carbanions are comparatively less comman. An example is given below:

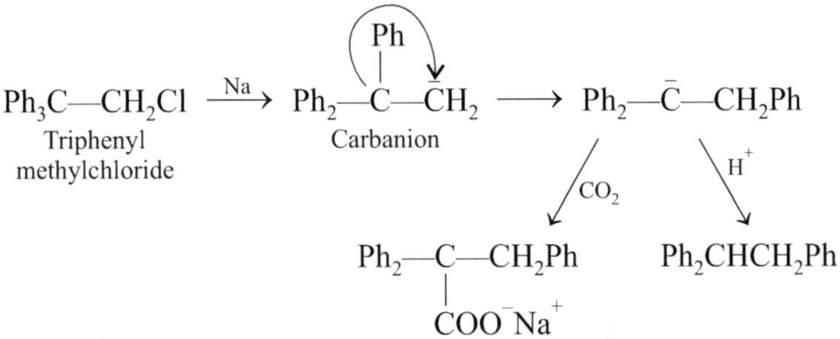

$$Ph_3C—CH_2Cl \xrightarrow{Na} Ph_2\overset{\overset{\displaystyle Ph}{|}}{—}C—\bar{C}H_2 \longrightarrow Ph_2—\bar{C}—CH_2Ph$$

Triphenyl methylchloride   Carbanion

$$Ph_2—C—CH_2Ph \qquad Ph_2CHCH_2Ph$$
$$\underset{\underset{\text{(Rearrangement of carbanions)}}{|}}{COO^-Na^+}$$

with CO_2 branch and H^+ branch leading to the above products.

Decarboxylation

The carboxylate anion lose CO_2 via the intermediate formation of carbanion followed by acquiring a proton from the solvent.

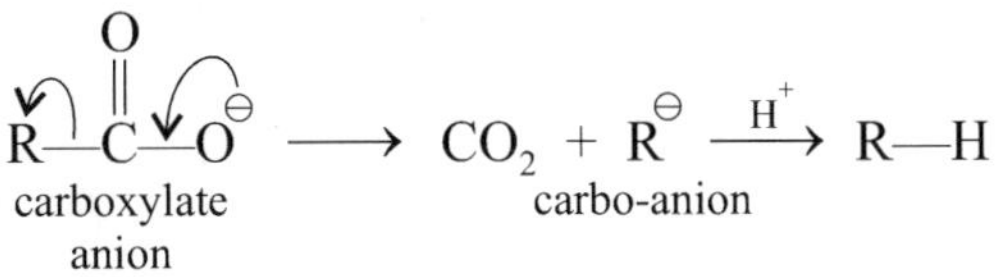

$$R—\overset{\overset{\displaystyle O}{||}}{C}—O^{\ominus} \longrightarrow CO_2 + R^{\ominus} \xrightarrow{H^+} R—H$$

carboxylate anion   carbo-anion

Oxidation of Carbanions

Under suitable conditions carbanions undergo oxidation. Thus, triphenyl methyl anion can be oxidised to triphenylmethyl radical, which can be reduced back to carbanion.

$$Ph_3C—\overset{\ominus}{C}H_2 \quad Ph\bar{C}Na^+ \underset{}{\overset{O_2}{\rightleftharpoons}} Ph_3\dot{C}$$

Triphenylmethyl radical

(Oxidation of carbanion)

In same case the carbanions can be oxidised with one electron oxidising agent like iodine to give coupled products from the formed radical. An example is given below:

$$(CH_3CO)_2\bar{C}H \xrightarrow{I_2} (CH_3CO)_2\dot{C}H$$

Anion from acetyl acetone

$$2(CH_3CO)_2\dot{C}H \longrightarrow (CH_3CO)_2—CH$$
$$|$$
$$(CH_3CO)_2—CH$$

coupled product

(oxidation of carbanion with iodine)

Reactions Involving Carbanions

A number of reactions involve the intermediate formation of carbanions. These include Aldol condensation, Perkin reaction, Claisen condensation, Dceckmann condensation and Michael addition

Aldol Condensation

As already stated in this reaction two molecules of an aldehyde react in presence of a base to give product called aldol

$$CH_3CHO + CH_3CHO \xrightarrow[\text{OH}]{} CH_3\overset{\overset{\displaystyle OH}{|}}{C}HCH_2CHO$$

$$\text{Acetaldehyde} \qquad\qquad \text{β-Hydroxy butyraldehyde}$$
$$\text{(aldol)}$$

The aldol reaction can take place between two identical or different aldehydes or ketones.

The mechanism of aldol condensation is given below:

Enolate anion
(resonance stabilized)

Acetaldehyde   Enolate anion   Alkoxide anion

Alkoxide anion   Aldol

Acid-Catalysed Aldol Condensation

In the aldol condensation cited above, the condensation takes place in presence of a base. However, aldol condensations can also be brought about with acid catalyst. For example, treatment of acetone with hydrogen chloride gives the aldol condensation product, *viz.*, 4-methyl-3-penten-2-one. In general, in acid-catalysed aldol reactions, there is simultaneous dehydration of the initially formed aldol.

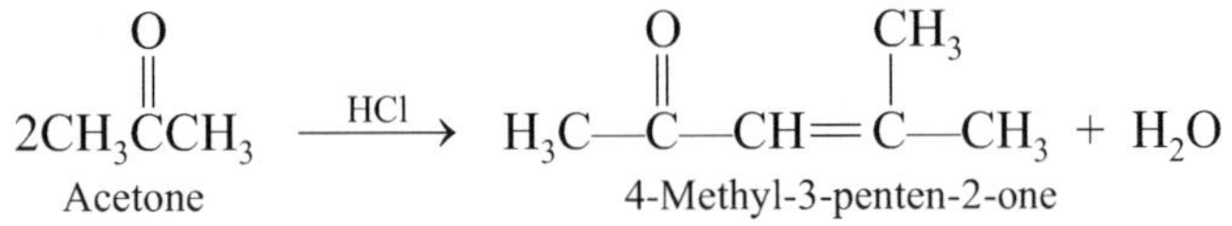

$$2CH_3\overset{\overset{\displaystyle O}{\|}}{C}CH_3 \xrightarrow{\text{HCl}} H_3C-\overset{\overset{\displaystyle O}{\|}}{C}-CH=\overset{\overset{\displaystyle CH_3}{|}}{C}-CH_3 + H_2O$$

Acetone   4-Methyl-3-penten-2-one

The mechanism of the acid-catalysed aldol condensation starts with the acid-catalysed formation of enol, which adds to the protonated carbonyl group of another molecule of acetone. The final step is proton transfer and dehydration leading to the final end product.

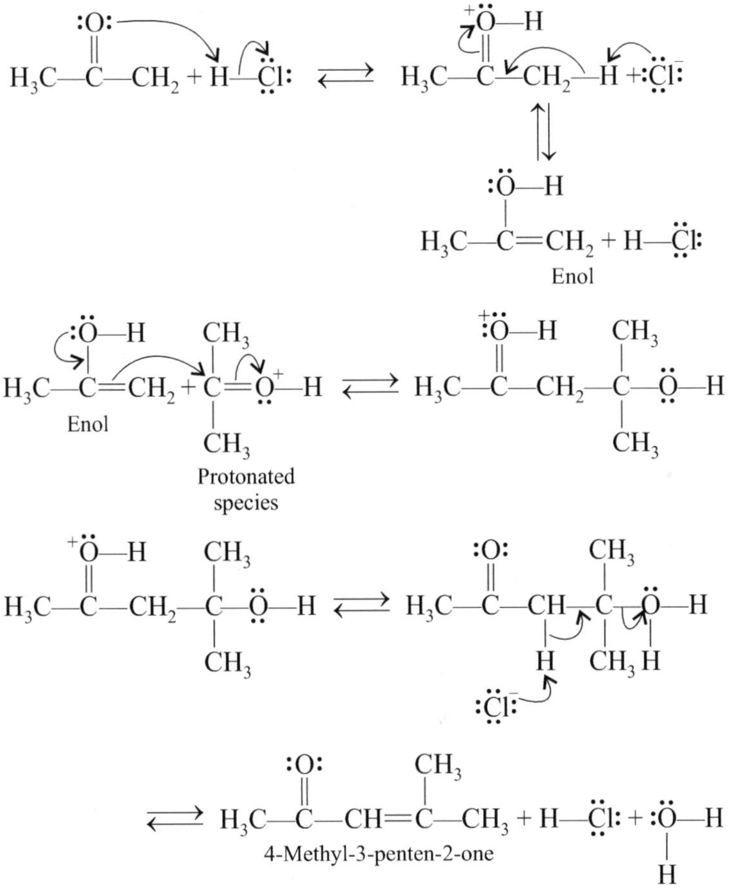

Enol

Enol

Protonated species

:Cl:⁻

4-Methyl-3-penten-2-one

Crossed Aldol Condensation

An aldol condensation that uses two different carbonyl compounds is called a crossed aldol condensation. In such a situation, the following three situations may be there:

(*i*) **Crossed aldol Condensation between two different aldehydes.** In case, both the aldehydes have α-hydrogen(s), both can form carbanions and so a mixture of four products are formed. Such a reaction has no synthetic utility.

If, on the other hand, one of the aldehydes has no α-hydrogen then in such a case two products are formed as shown below

$$(a) \quad R_3C\text{—}CHO + CH_3CHO \xrightarrow{\ ^-OH\ } R_3C\overset{\overset{\displaystyle OH}{|}}{\text{—}CH}\text{—}CH_2CHO$$

(Crossed product)

$$(b) \quad CH_3CHO + CH_3CHO \xrightarrow{\ ^-OH\ } H_3C\overset{\overset{\displaystyle OH}{|}}{\text{—}CH}\text{—}CH_2CHO$$

(Normal simple product)

Perkin Reaction

It involves the reaction of an aldehyde (say benzaldehyde) with acetic anhydride in the presence of sodium acetate. Acetic anhydride first reacts with sodium acetate when H of an α carbon atom from the molecule of acetic anhydride is removed (the anion of sodium acetate acts here as a base). The carbanion then reacts with benzaldehyde, giving cinnamic acid as the final product.

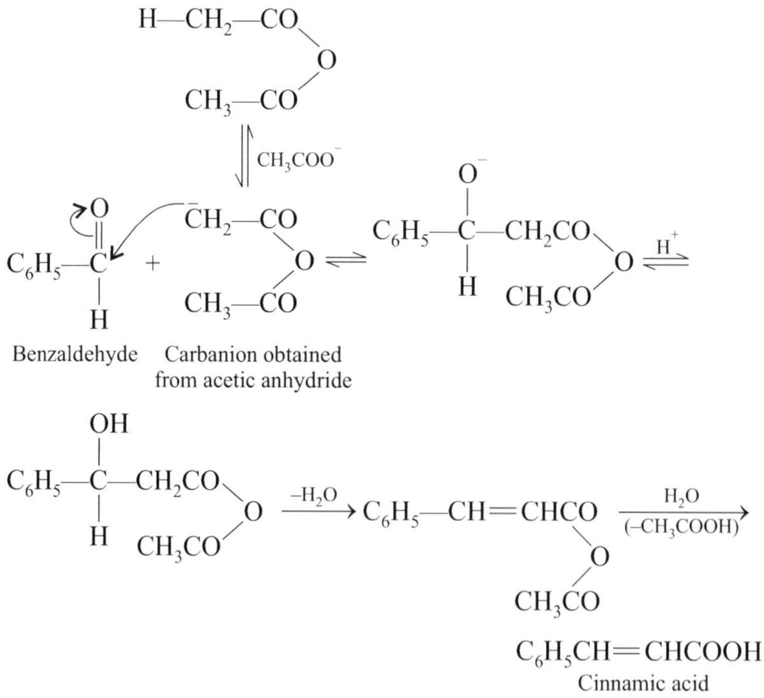

Benzaldehyde   Carbanion obtained from acetic anhydride

$$C_6H_5\text{—}CH\text{=}CHCOOH$$

Cinnamic acid

Claisen condensation. In this reaction, the carbanion derived from an ester molecule (say ethyl acetate) reacts with another molecule of the same ester to give a β-keto ester (ethyl acetoacetate) as the possible product.

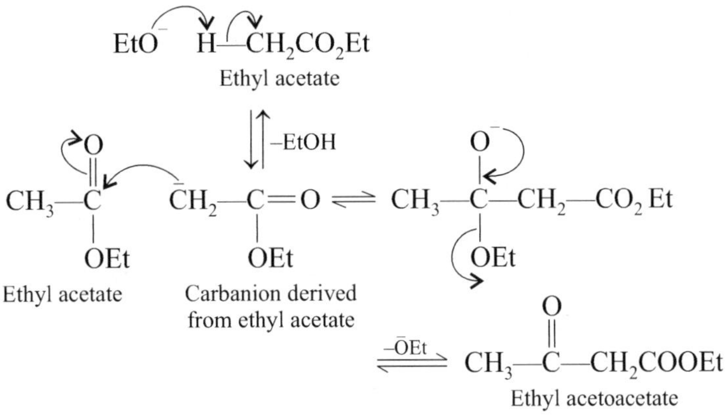

EtO H—CH$_2$CO$_2$Et
Ethyl acetate

CH$_3$—C(OEt)=O + CH$_2$—C=O(OEt) $\rightleftharpoons$ CH$_3$—C—CH$_2$—CO$_2$Et(OEt)

$\downarrow\uparrow$ –EtOH

Ethyl acetate   Carbanion derived
                       from ethyl acetate

$\xrightarrow{\text{–OEt}}$ CH$_3$—C—CH$_2$COOEt (O)

Ethyl acetoacetate

Dieckmann condensation. It may be considered as an intermolecular Claisen condensation and is quite useful in the preparation of cyclic ketones as the final products. Diesters of C$_6$ and C$_7$ dibasic acids give good yields of cyclic β-ketoesters. Thus, ethyl esters of adipic acid and pimelic acid give 2-carbethoxycyclopentanone and 2-carbethoxycyclohexanone, respectively.

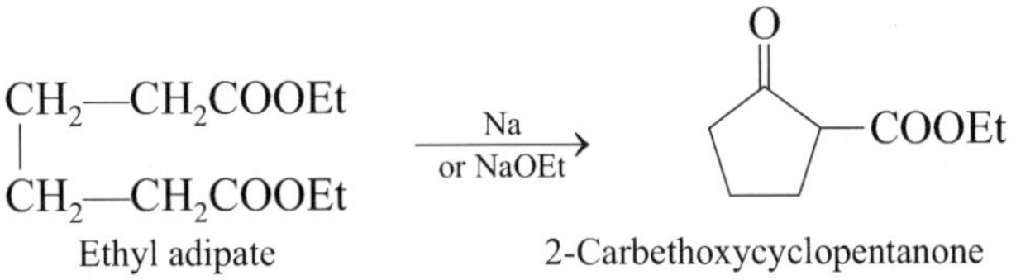

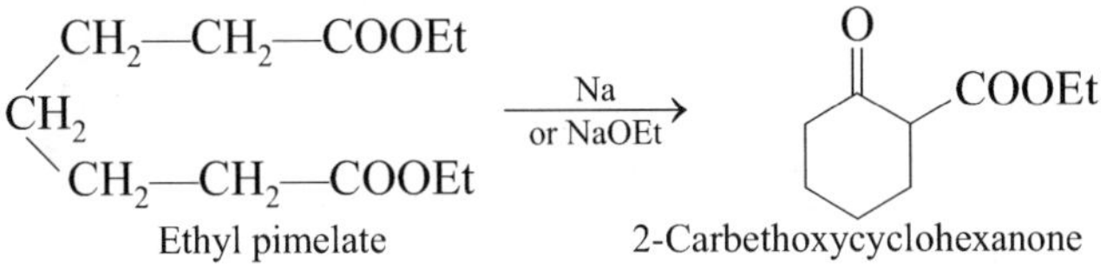

CH$_2$—CH$_2$COOEt
|
CH$_2$—CH$_2$COOEt
Ethyl adipate

$\xrightarrow[\text{or NaOEt}]{\text{Na}}$

2-Carbethoxycyclopentanone — COOEt

CH$_2$—CH$_2$—COOEt
CH$_2$
CH$_2$—CH$_2$—COOEt
Ethyl pimelate

$\xrightarrow[\text{or NaOEt}]{\text{Na}}$

2-Carbethoxycyclohexanone COOEt

Mechanism of the reactions:

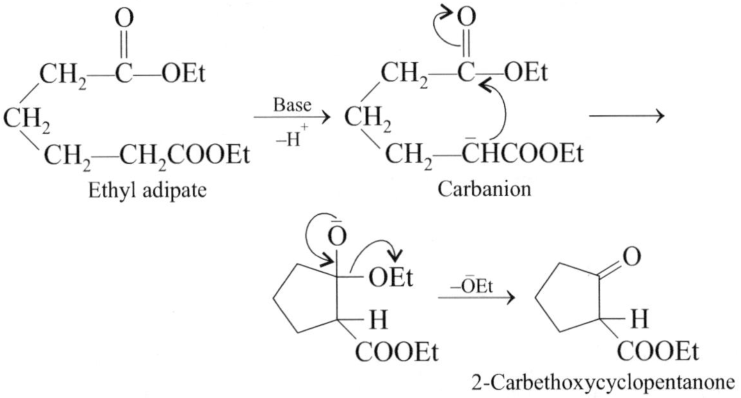

CH$_2$—C—OEt (O)
CH$_2$
CH$_2$—CH$_2$COOEt
Ethyl adipate

$\xrightarrow[\text{–H}^+]{\text{Base}}$

CH$_2$—C—OEt (O)
CH$_2$
CH$_2$—CHCOOEt
Carbanion

$\longrightarrow$

OEt
—H
COOEt

$\xrightarrow{\text{–OEt}}$

O
—H
COOEt

2-Carbethoxycyclopentanone

Michael addition. It is addition reaction between an α, β-unsaturated carbonyl compound and a compound with active methylene group, *e.g.*, malonic ester, acetoacetic ester, cyanoacetic ester and nitroparaffins in presence of a base *e.g.*, sodium ethoxide, or a secondary amine (usually piperidine). The reaction is illustrated as given below:

$$CH_2=CH-\overset{\overset{\displaystyle O}{||}}{C}-CH_3 + CH_2(COOC_2H_5)_2 \xrightarrow{\text{NaOEt}}$$

$$(H_5C_2OOC)_2CHCH_2CH_2-\overset{\overset{\displaystyle O}{||}}{C}-CH_3$$

Mechanism of the reactions:

$$CH_2(COOC_2H_5)_2 + \bar{O}C_2H_5 \longrightarrow \bar{C}H(COOC_2H_5)_2 + C_2H_5OH$$

Malonic ester   Carbanion

$$\bar{C}H(COOC_2H_5)_2 + H_2C=CH-\overset{\overset{\displaystyle O}{||}}{C}-CH_3 \rightleftharpoons$$

Carbanion   α,β-unsaturated carbonyl compound

$$H_2C-CH=\overset{\overset{\displaystyle \bar{O}}{|}}{C}-CH_3 \underset{C_2H_5OH}{\rightleftharpoons} H_2C-\overset{|}{C}H=\overset{\overset{\displaystyle O-H}{|}}{C}-CH_3 \rightleftharpoons$$

$$\underset{CH(COOC_2H_5)_2}{|} \qquad\qquad \underset{CH(COOC_2H_5)_2}{|}$$

$$H_2C-CH_2-\overset{\overset{\displaystyle O}{||}}{C}-CH_3$$

$$\underset{CH(COOC_2H_5)_2}{|}$$

Adduct

2.4.3.3 Free Radicals

We have know that when a covalent bond undergoes homolytic fission, the two departing atoms take one of the bonding pair of electrons each.

$$A : B \xrightarrow[\text{fission}]{\text{Homolytic}} \overset{.}{A} + \overset{.}{B}$$

The two frangments produced carry an odd electron each and are called *free radicals.* These free radicals are reaction intermediates and are very reactive and transitory existance. These react with other readicals or molecules by gaining one more electron in order to restore the stable bonding pair. The high reactivity of free radicals is due to the tendency of the odd electron to 'pair-up' with aother available electron.

In fact, the term free radical is used for any species which possess an unpaired electron. Free radicals are the most important reaction intermediates.

A free radicals is paramagnetic and so can be observed by electron spin resonance (ESR) spectroscopy. For simple *free radicals,* two possible structures have been postulated. The first is a planar sp^2 hybridised radical (A), similar to carbocation. The second is pyramidal sp^3 hybridised radical (B), similar to carbanion.

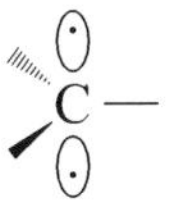

(A)
Planar

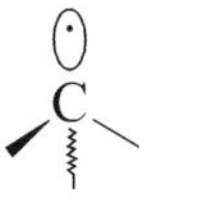

(B)
Pyramidal

There is no chemical evidence supporting either a planar structure (A) or a pyramidal structure (B) for simple free radicals. However, physical methods like ESR, UV and IR, indicate that simple free radicals have the planar structure (A).

Stability of Free Radicals

Like carbocations, the stability of the free radicals is also of the order:

tertiary > secondary > primary.

The stability can be explained by hyperconjugation as in the case of carbocations. The stability of free radicals also depends on the resonance. Thus, allylic and benzylic free radicals are more stable than simple alkyl radicals. This is because of the delocatization of the unpaired electron over the π orbital system in each case.

The stability of a radical increases as the extent of delocatization increases. Thus, $Ph_2\dot{C}H$ is more stable than $Ph\dot{C}H_2$ and $Ph_3\dot{C}$ is a reasonably stable radical. The stability of $Ph_3\dot{C}$ can be explained by the delocalization of the unpaired electron. In case of $Ph_3\dot{C}$, the delocalisation is supported by its e.s.r spectrum.

The extent of delocalization is maximum in $Ph_3\dot{C}$ than in $Ph_2\dot{C}H$ or even in $Ph\dot{C}H_2$

$$RCH{=}CH\,\dot{C}H_2 \longleftrightarrow \left[RCH{=\!=}CH{=\!=}CH_2\right]^{\displaystyle\cdot}$$

Allyl free
radical

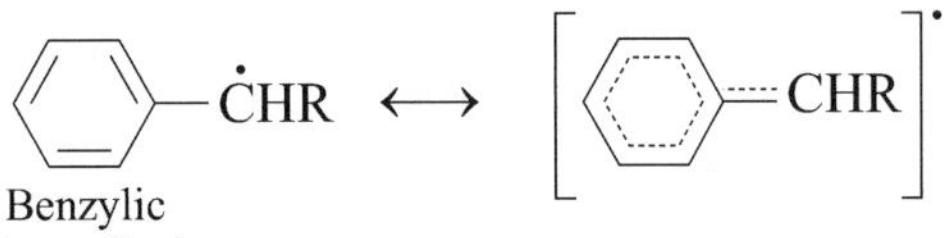

Benzylic
free radical

$Ph_3\dot{C}$
Triphenyl methyl
radical

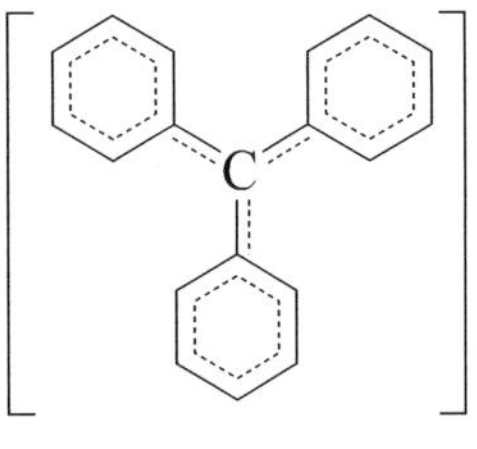

On the basis of spectroscopic and X-ray crystallographic studies, triarylmethyl radicals have been shown to be propeller shaped (as shown below), the benzene rings being angled at about 30° out of common plane.

Triarylmethyl radical
(propeller-shaped)

Some free radicals, known as **bridgehead free radicals** have fixed bond angles and dihedral angles. These have pyramidal structures. Their structures are supported on the basis of physical and chemical evidence. Following are given structures of some bridgehead radicals, which are conveniently formed.

(Bridged head radicals)

The radicals described so far are carbon radicals. Besides these we come across other radicals containing nitrogen, sulphure and oxygen. Such radicals are called **hetero radicals.** These have varing degrees of stability.

A nitrogen radical is obtained by warming N, N, N′, N′-tetraphenyl hydrazine (which is obtained by the oxidation of diphenyl amine with KMnO$_4$ in a non-polar solvent). The solution acquires green colour due to the formation of hetero radical, ph$_2$ $\dot{N}$.

$$2Ph_2NH \xrightarrow[MnO_4^{\ominus}]{[O]} Ph_2N-NPh_2 \underset{}{\overset{CCl_4. \text{ Warm}}{\rightleftharpoons}} Ph_2\dot{N} + \dot{N}Ph$$

Diphenyl amine   N, N, N′,N′- tetraphenyl hydrazine   Green colour Nitrogen radical

Another nitrogen radical, 1, 1-diphenyl-2-picrylhydrazyl is obtained by the oxidation of triarylhydrazine with PbO$_2$

$$Ph_2NNH_2 \xrightarrow[\text{chloride}]{\text{Picryl}} Ph_2NNH-\overset{O_2N}{\underset{O_2N}{\bigcirc}}-NO_2 \xrightarrow[PbO_2]{[O]} Ph_2N\dot{N}-\overset{NO_2}{\underset{NO_2}{\bigcirc}}-NO_2$$

Diphenyl hydrazine

1,1-Diphenyl-2-picryl hydrazyl radical
(stable, violet colour)

Thiyl radicals are sulphur containing radicals such as $Ph\dot{S}$. These are obtained by heating diphenyl disulphide. This radical is yellow coloured in solid state and becomes colourless on cooling.

$$PhS\text{—}SPh \underset{}{\overset{\Delta}{\rightleftharpoons}} PhS\cdot + \cdot SPh$$

Diphenyl disulphide   Thiyl radicals

Phenoxy radical are oxygen containing radicals and are relatively stable. These are obtained by the oxidation of appropriately substituted phenoxide with $K_3Fe(CN)_6$.

2,4,6-Tert-butyl phenoxide   2,4,6-Tert-butyl phenoxy radical Dark blue solid (m.p. 97°C)

It is a relatively unreactive radical due to steric hinderence by the bulky CMe_3 groups in both the ortho positions.

Generation of Free Radicals

Following are given some of the methods used for the generation of free radical. These include photolysis, thermolysis and redox reactions.

> (*i*) **Photolysis.** Only compounds, which absorb in the ultra-violet or visible range undergo photolysis. Thus, acetone in the vapour phase undergoes photolysis by light of wave length 320 nm (3200 Å) to give two molecules of methyl free radicals

$$Me\text{—}\overset{\overset{\displaystyle O}{\|}}{C}\text{—}Me \xrightarrow{h\nu} Me\cdot + \cdot\overset{\overset{\displaystyle O}{\|}}{C}\text{—}Me$$
$$\longrightarrow CO + \cdot Me$$

Other species like hypochlorites and alkyl nitrites also undergo photolysis to give alkoxy radicals

$$R\text{—}O\text{—}Cl \xrightarrow{h\nu} RO^{\cdot} + \cdot Cl$$
Alkyl hypochlorite   Alkoxy radical

$$R\text{—}O\text{—}NO \xrightarrow{h\nu} RO\cdot + \cdot NO$$
Alkyl nitrite   Alkoxy radical

Halogens also undergo photolytic homolysis to yield free radicals

$$\underset{\text{Chlorine}}{Cl\text{—}Cl} \xrightarrow{h\nu} Cl\cdot + \cdot Cl$$

$$\underset{\text{Bromine}}{Br\text{—}Br} \xrightarrow{h\nu} Br\cdot + \cdot Br$$

The halogen free radicals (Cl· or Br·) obtained above can initiate halogenation of alkanes or addition to alkenes.

Photolysis can cleave even strong bonds that do not readily break. Thus, azoalkanes on photolysis give alkyl free radicals.

$$\underset{\text{Azoalkanes}}{R\text{—}N=N\text{—}R} \xrightarrow{h\nu} R\cdot + N\equiv N + \cdot R$$

In photolysis, energy of only one particular wave length is transferred to a molecule. So this is a more specific method of effecting homolysis then in pyrolysis. As an example, cleavage of diacetyl peroxide by photolysis generate the free radicals, compared to the formation of a number of products by thermolysis

(*ii*) **Thermolysis.** This method involved heating an organic substrate at suitable temperature. Some example are given below:

The benzyl and tert. butyl free radicals find application in polymers.

(*iii*) **Redox Reactions.** Redox reactions involve one electron transfer for the generation of free radicals. The one-electron source is the metal ion (*e.g.*, Cu^+, Fe^{2+}). As an example Cu^+ ions are used for the decomposition of acyl peroxides.

$$Ar-\overset{\overset{\displaystyle O}{\|}}{C}-O-O-\overset{\overset{\displaystyle O}{\|}}{C}-Ar + Cu^+ \longrightarrow Ar-\overset{\overset{\displaystyle O}{\|}}{C}-O^{\bullet} + ArCO_2^- + Cu^{2+}$$
Acylperoxide   Acyloxy free
radical

This is a convenient method for the generation of Acyloxy free radical

$(Ar-\overset{\overset{\displaystyle O}{\|}}{C}-O^{\bullet})$, since in thermolysis, the acyloxy radical decompose to give $Ar. + CO_2$.

Iron in the Ferrous state (Fe^{2+}) is used to catalyse the oxidation of aqueous H_2O_2.

$$H_2O_2 + Fe^{2+} \longrightarrow HO^{\bullet} + {}^-OH + Fe^{3+}$$

Fenton's reagent is a mixture of H_2O_2 and Fe^{2+} the effective oxidizing agent is hydroxy radical (HO.) is used to generate another free radical, which may dimerise

$$H\overset{\bullet}{O} + H-CH_2-C\,Me_2OH \longrightarrow H_2O + \cdot CH_2CMe_2OH$$
Hydroxy   Trimethyl carbinol   Free radical
radical

$\downarrow$ Dimeroe

$$HOMe_2CCH_2CH_2CMe_2OH$$
Dimeric product

The auto oxidation of benzaldehyde to benzoyl free radical is also catalysed by one electron transfer.

$$Ph-\overset{\overset{\displaystyle O}{\|}}{C}-H + Fe^{3+} \longrightarrow Ph\,\overset{\overset{\displaystyle O}{\|}}{C}{}^{\bullet} + H^{\oplus} + Fe^{2+}$$
Benzaldehyde

The generation of a stable phenoxy radical (also taken place by one-electron oxidation by $Fe(CN)_6$.

$$+ Fe(CN)_6^{3-} \longrightarrow$$ $$+ Fe(CN)_6^{4-}$$

Radicals, which subsequently dimense are also obtained by anodic oxidation of carboxylate anions, RCO_2^-, in the Kolbe electrolytic synthesis of hydrocarbons.

$$2RCO_2^- \xrightarrow[\text{anode}]{-e^-} 2RCO^{\bullet} \xrightarrow{-CO_2} 2R^{\bullet} \longrightarrow R-R$$

Reactions of Free Radicals

Free radicals undergo a number of reactions like recombination, dis-proportionation, reaction with iodine, reaction with iodine and metals and rearrangement.

(*i*) **Recombination of Free Radicals.** Being very reactive reaction intermediates, the radicals may recombine to form hydrocarbon

$$\dot{C}H_3 + \dot{C}H_3 \longrightarrow CH_3—CH_3$$
Methyl free radical   Ethane

$$2CH_3\dot{C}H_2 \longrightarrow CH_3CH_2CH_2CH_3$$
Ethyl free radical   *n*-Butane

The recombination of free radicals is used in the termination step in free radical polymerisation for the manufacture of polymers.

(*ii*) **Disproportionation.** Alkyl radicals undergo disproportionation at higher temperatures. Thus, ethyl free radical disproportionates to gives ethylene and ethane. In this process an radical $CH_3\,\dot{C}H_2$ takes up a hydrogen from another free radical.

$$CH_3\dot{C}H_2 + CH_3\dot{C}H_2 \longrightarrow CH_2{=}CH_2 + CH_3—CH_3$$
Ethyl Free   Ethylene   Ethane
radical

(*iii*) **Reaction with Olefins.** The alkyl free radicals reacts with olefins to generate other free radical, which may further react with another molecule of olefin to generate another free radical. The reaction continues till the formed free radical couples with another free radical and the reaction gets terminated.

$$\dot{C}H_3 + CH_2{=}CH_2 \longrightarrow CH_3—CH_2—\dot{C}H_2$$

$$CH_3—CH_2—\dot{C}H_2 + CH_2{=}CH_2 \longrightarrow CH_3CH_2CH_2CH_2\dot{C}H_2$$

(*iv*) **Reaction with iodine and metals.** The alkyl radicals may combine with iodine and other elements to form alkyl derivatives. Some examples include:

$$2\dot{C}H_3 + I_2 \longrightarrow 2CH_3I$$

$$2\dot{C}H_3 + Zn \longrightarrow (CH_3)_2Zn$$
Dimethyl Zinc

$$2\dot{C}H_3 + Hg \longrightarrow (CH_3)_2Hg$$
Dimethyl mercury

(*v*) **Rearrangements.** Like carbocations and carbanions, the free radicals also undergo rearrangements involving 1, 2-shifts of aryl group via. a bridged transition state. Thus $Me_3\dot{C}O$ radical abstracts a proton from the CHO group to give an acyl radical, which loses CO and forms another radical. This radical in turn undergoes rearrangement to give another radical. In this case the rearrangement involves migration of Ph via the bridged ion intermediate.

Acyl radical

Ph Me $\dot{C}$—CH$_2$
Rearranged radical

Ph Me C——CH$_2$
Bridged ion intermediate

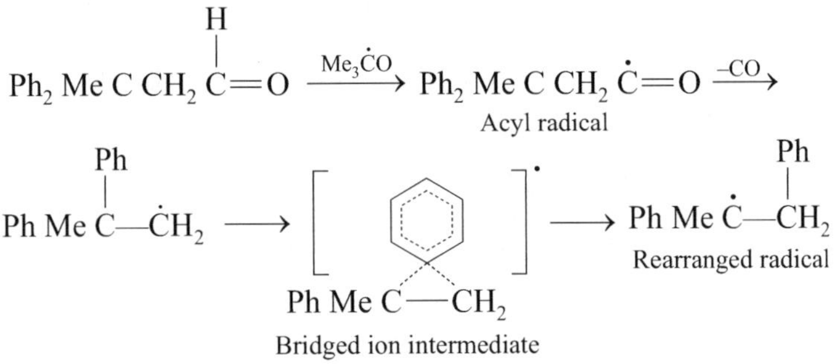

Reactions Involving Free Radicals

A number of reactions proceed via the formation of free radicals. These include Sandmeyer reaction, Gomberg reaction, Wurtz reaction, Hunsdiecker reaction, Kolbes electrolytic reaction, halogenation and vinyl polymensation

Sandmeyer reaction. It involves the copper (I) catalysed decomposition of diazonium salts for the preparation of benzene derivatives.

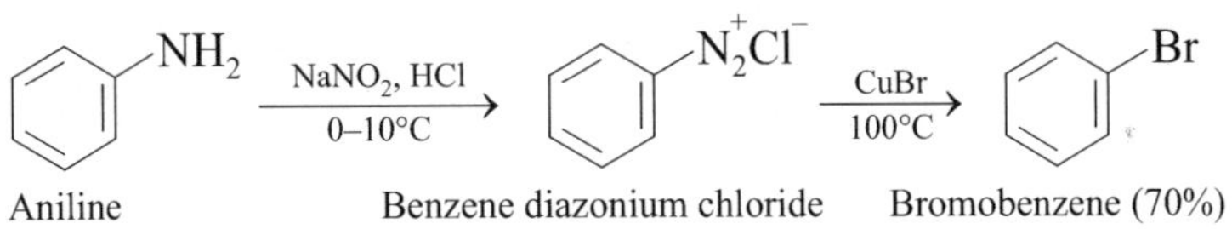

Aniline Benzene diazonium chloride Bromobenzene (70%)

In the final step of this reaction, a free radical and cupric ions are produced and the reaction is completed as follows:

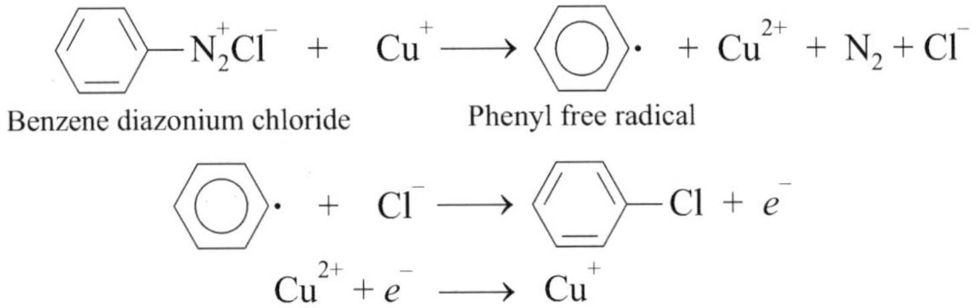

Benzene diazonium chloride Phenyl free radical

$$Cu^{2+} + e^- \longrightarrow Cu^+$$

Gomberg reaction. This reaction is used to prepare monosubstituted biphenyls by the reaction of an alkaline solution of a diazonium salt with an aromatic compound (say benzene). For example,

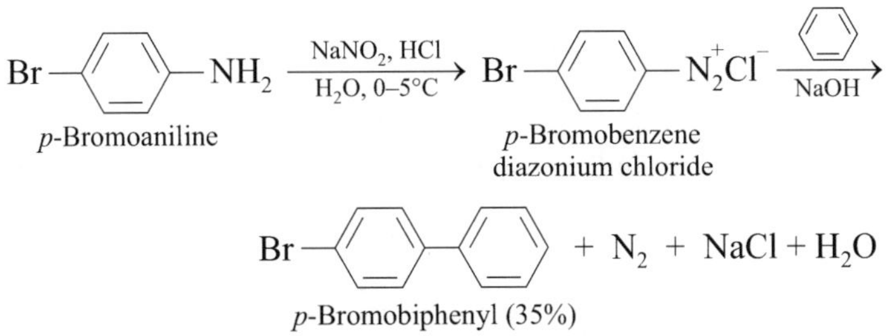

p-Bromoaniline *p*-Bromobenzene diazonium chloride

p-Bromobiphenyl (35%)

The reaction proceeds via the formation of a free radical as explained below:

$$C_6H_5N_2^+Cl^- \xrightarrow{\text{OH}^-} C_6H_5{-}N{=}N{-}OH \longrightarrow C_6\overset{.}{H_5} + N_2 + \overset{.}{O}H$$

$$C_6\overset{.}{H_5} + C_6H_5NO_2 + \overset{.}{O}H \longrightarrow p\text{-}C_6H_5{-}C_6H_4{-}NO_2 + H_2O$$

Wurtz reaction. Coupling of alkyl halides with sodium in dry ether to give hydrocarbons is known as Wurtz reaction.

$$\underset{\text{Propyl bromide}}{CH_3CH_2CH_2Br} \xrightarrow[\text{Ether}]{\text{Na}} \underset{\text{Hexane}}{CH_3(CH_2)_4CH_3}$$

This reaction is especially useful for the synthesis of alkanes containing even number of carbon atoms (*i.e.*,symmetrical alkanes).

A *radical mechanism*, involving the following steps, has been proposed for this reaction.

$$RX + Na \longrightarrow \underset{\text{Free radical}}{\overset{.}{R}} + NaX$$

$$\overset{.}{R} + \overset{.}{R} \longrightarrow R{-}R$$

Hunsdiecker reaction. This reaction is useful for the preparation of organic halides (alkyl or aryl halides) by the thermal decomposition of silver salt of carboxylic acids in the presence of a halogen. This reaction is described as decarboxylative halogenation.

$$RCOOAg + X_2 \xrightarrow{\Delta} RX + CO_2 + AgX$$

Mechanism of the reaction:

$$\underset{}{R{-}\overset{\overset{\displaystyle O}{\|}}{C}{-}OAg} + Br_2 \longrightarrow \underset{\text{Acylhypohalite}}{R{-}\overset{\overset{\displaystyle O}{\|}}{C}{-}OBr} + AgBr$$

$$\underset{}{R{-}\overset{\overset{\displaystyle O}{\|}}{C}{-}OBr} \longrightarrow \underset{\text{Acyloxy radical}}{R{-}\overset{\overset{\displaystyle O}{\|}}{C}{-}\overset{.}{O}} + \overset{.}{Br}$$

$$R{-}\overset{\overset{\displaystyle O}{\|}}{C}{-}\overset{.}{O} \xrightarrow{-CO_2} \underset{\text{Alkyl radical}}{\overset{.}{R}}$$

$$\overset{.}{R} + RCOOBr \longrightarrow \underset{\text{Alkyl bromide}}{RBr} + R{-}\overset{\overset{\displaystyle O}{\|}}{C}{-}\overset{.}{O}$$

Kolbe electrolytic reaction. This reaction is used for the preparation of hydrocarbons by the electrolysis of sodium or potassium salts of carboxylic acids.

$$2RCOO^- \xrightarrow{\text{Electrolysis}} R{-}R + 2CO_2$$

A *free radical mechanism* is proposed for this reaction. The anodic oxidation of carboxylate anion gives alkyl radical via an alkoxy radical. The alkyl radicals then dimerise to give alkane molecule.

$$
\underset{\text{Carboxylate anion}}{R-\overset{\overset{\displaystyle \cdot O}{\|}}{C}-O^{-}} \xrightarrow{-1e^{-}} \underset{\text{Alkoxy radical}}{R-\overset{\overset{\displaystyle O}{\|}}{C}-\overset{\cdot}{O}} \xrightarrow{-CO_2} \underset{\text{Alkyl radical}}{\overset{\cdot}{R}}
$$

$$
2\,\overset{\cdot}{R} \longrightarrow \underset{\text{Alkane}}{R-R}
$$

Halogenation

In most of the cases, halogenation proceeds vier the formation of free radicals:

Some typical halogenation include

- Conversion of methane into carbon tetrachloride
- Conversion of benzene into benzene hexachloride
- Conversion of toluene into benzyl chloride
- Conversion of propene into *n*-propyl bromide

(*i*) **Conversion of methane into carbon tetrachloride.** Treatment of methane with excess of chlorine in presence of sun light gives carbon tetrachloride.

$$
CH_4 + \underset{\text{excess}}{Cl_2} \xrightarrow{hv} CCl_4 + HCl
$$

The reaction proceeds via the formation of free radicals

(*i*) $:\ddot{\underset{..}{C}l\!:\;\;\ddot{\underset{..}{C}}l: \longrightarrow :\overset{..}{\underset{..}{C}}l\cdot \;+\; \cdot\overset{..}{\underset{..}{C}}l:}$ (Initiation step)

Chlorine free radical

(*ii*) $:\overset{..}{\underset{..}{C}}l\cdot \;+\; H\!:\!\overset{\overset{\displaystyle H}{}}{\underset{\underset{\displaystyle H}{}}{C}}\!:\!H \longrightarrow \overset{\overset{\displaystyle H}{\diagdown}}{\underset{\underset{\displaystyle H}{\diagup}}{C}}\!-\!H$ (Propagation step)

Methyl free
radical

(*iii*) $H-\overset{\overset{\displaystyle H}{|}}{\underset{\underset{\displaystyle H}{|}}{C}}\cdot \;+\; :\overset{..}{\underset{..}{C}}l\!:\!\overset{..}{\underset{..}{C}}l: \longrightarrow H-\overset{\overset{\displaystyle H}{|}}{\underset{\underset{\displaystyle H}{|}}{C}}-Cl \;+\; \cdot\overset{..}{\underset{..}{C}}l:$

Methyl chloride

Steps (*ii*) and (*iii*) are repeated to finally give carbon tetrachloride. The reaction is finally terminated by any of the following three steps

$$
\left.
\begin{array}{l}
Cl\cdot + Cl\cdot \longrightarrow Cl_2 \\
Cl\cdot + \cdot CH_3 \longrightarrow CH_3-Cl \\
\cdot CH_3 + \cdot CH_3 \longrightarrow CH_3-CH_3
\end{array}
\right]
\quad \text{Termination step}
$$

(*ii*) **Conversion of benzene into benzene hexachloride.** The reaction of benzene with chlorine proceeds via a free radical pathway. The reaction is carried out in presence of light or peroxide.

$$Cl_2 \xrightarrow{h\nu} 2\dot{C}l$$

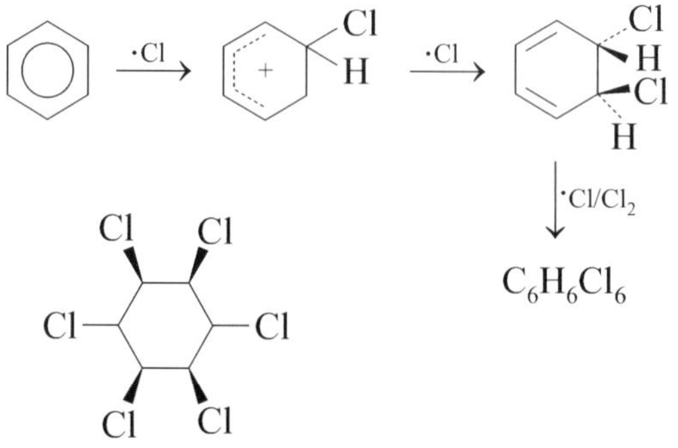

Lindane

The above reaction gives a mixture of eight non-converteble stereoisomers of hexachlorobenzene. The gamma isomer is known as Lindane (gammexane) and is a useful insecticide.

(*iii*) **Conversion of toluene into benzyl chloride.** Toluene reacts with chlorine in presence of sunlight to give benzyl chloride. The chlorine radical, $\dot{C}l$ (formed from chlorine in presence of sunlight) abstracts a proton to give delocalized benzyl radical, $C_6H_5\dot{C}H_2$ (which is resonance stabilized) rather than hexadecyl radical, in which the aromatic stabilization of the starting material is lost

$$Cl_2 \xrightarrow{h\nu} 2\dot{C}l$$

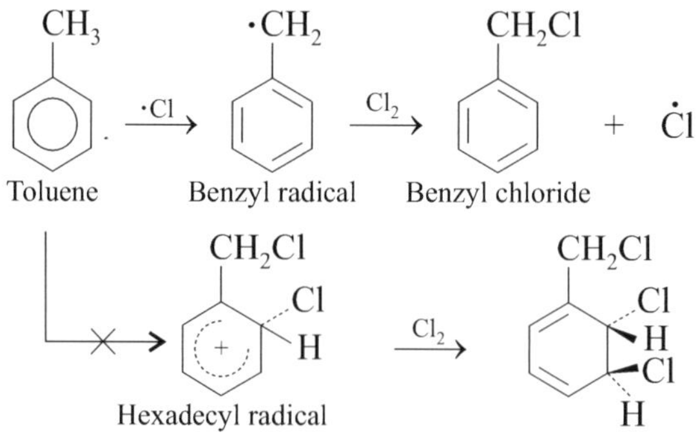

| Toluene | Benzyl radical | Benzyl chloride |

Hexadecyl radical

(*iv*) **Conversion of propene into *n*-propyl bromide.** The addition of HBr to propene is known to give isopropyl bromide. In this case, the reaction proceeds via a carbocation intermediate. The initially formed primary carbocation being unstable gets converted into more stable secondary carbocation. The reaction takes place in accordance with **Markownikoff rule** and is represented as give below:

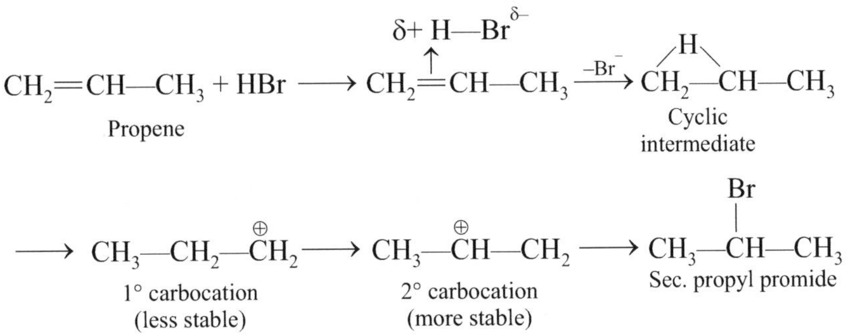

$$\delta+ \text{H}—\text{Br}^{\delta-}$$

$$\text{CH}_2\!\!=\!\!\text{CH}—\text{CH}_3 + \text{HBr} \longrightarrow \text{CH}_2\!\overset{\uparrow}{=}\!\text{CH}—\text{CH}_3 \xrightarrow{-\text{Br}^-} \text{CH}_2\!\!\overset{\diagup \text{H} \diagdown}{—}\!\text{CH}—\text{CH}_3$$

Propene

Cyclic
intermediate

$$\longrightarrow \text{CH}_3—\text{CH}_2—\overset{\oplus}{\text{CH}}_2 \longrightarrow \text{CH}_3—\overset{\oplus}{\text{CH}}—\text{CH}_2 \longrightarrow \text{CH}_3—\overset{\overset{\text{Br}}{|}}{\text{CH}}—\text{CH}_3$$

1° carbocation
(less stable)

2° carbocation
(more stable)

Sec. propyl promide

In case the above reaction is carried out in presence of peroxide, the product obtained is *n*-propyl bromide. This is known as **peroxide effect** and the reaction takes place in anti-markownikoff fashion. In this case, the reaction proceeds by mechanism as shown below:

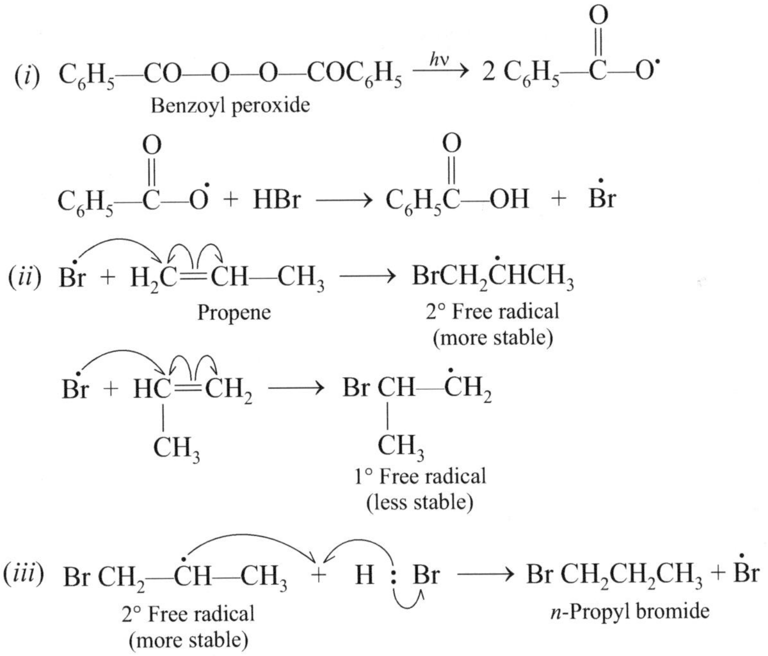

(*i*) $\text{C}_6\text{H}_5—\text{CO}—\text{O}—\text{O}—\text{COC}_6\text{H}_5 \xrightarrow{hv} 2\,\text{C}_6\text{H}_5—\overset{\overset{\text{O}}{\|}}{\text{C}}—\text{O}^{\bullet}$

Benzoyl peroxide

$$\text{C}_6\text{H}_5—\overset{\overset{\text{O}}{\|}}{\text{C}}—\overset{\bullet}{\text{O}} + \text{HBr} \longrightarrow \text{C}_6\text{H}_5\overset{\overset{\text{O}}{\|}}{\text{C}}—\text{OH} + \overset{\bullet}{\text{Br}}$$

(*ii*) $\overset{\bullet}{\text{Br}} + \text{H}_2\text{C}\!\!=\!\!\text{CH}—\text{CH}_3 \longrightarrow \text{BrCH}_2\overset{\bullet}{\text{C}}\text{HCH}_3$

Propene

2° Free radical
(more stable)

$$\overset{\bullet}{\text{Br}} + \underset{\text{CH}_3}{\text{HC}\!\!=\!\!\text{CH}_2} \longrightarrow \underset{\text{CH}_3}{\text{Br CH}—\overset{\bullet}{\text{CH}}_2}$$

1° Free radical
(less stable)

(*iii*) $\text{Br CH}_2—\overset{\bullet}{\text{CH}}—\text{CH}_3 + \text{H} \!:\! \text{Br} \longrightarrow \text{Br CH}_2\text{CH}_2\text{CH}_3 + \overset{\bullet}{\text{Br}}$

2° Free radical
(more stable)

n-Propyl bromide

(*v*) **Vinyl Polymerisation.** The Vinyl polymerisation takes place by free radical mechanism and is used for the manufacture of polyethylene and polyvinyl chloride (PVC). This polymerisation takes place in three steps viz., initiation, propagation and termination.

Step (*i*) **Initiation.** As already stated, the reaction is initiated by free radical initiator like benzoyl peroxide. The free radical is generated as follows:

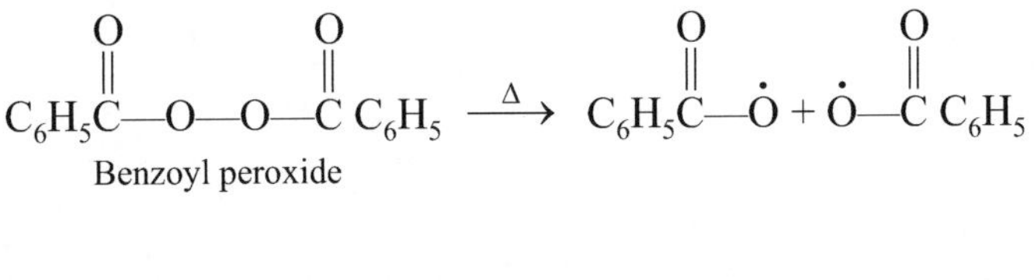

$$\text{C}_6\text{H}_5\overset{\overset{\text{O}}{\|}}{\text{C}}—\text{O}—\text{O}—\overset{\overset{\text{O}}{\|}}{\text{C}}\,\text{C}_6\text{H}_5 \xrightarrow{\Delta} \text{C}_6\text{H}_5\overset{\overset{\text{O}}{\|}}{\text{C}}—\overset{\bullet}{\text{O}} + \overset{\bullet}{\text{O}}—\overset{\overset{\text{O}}{\|}}{\text{C}}\,\text{C}_6\text{H}_5$$

Benzoyl peroxide

$$C_6H_5C-\overset{\overset{\displaystyle O}{\|}}{C}-O\cdot \xrightarrow{\Delta} C_6H_5^{\cdot} + CO_2$$

Phenyl
free radical

The free radical so formed adds on to a molecule of the monomer (ethylene or vinyl chloride).

$$R + CH_2=\overset{|}{\underset{X}{CH}} \longrightarrow R\,CH_2-\overset{\cdot}{\underset{X}{CH}}$$

Ethylene X = H   New free radical
Vinyl chloride X = Cl

Step (*ii*) **Propagation.** The new free radical generated in step (*i*) adds on to another molecule of the monomer to give another free radical. Successive addition of the so generated free radical to the monomer takes place.

$$R'\,CH_2-\overset{\cdot}{\underset{X}{CH}} \quad CH_2=\overset{|}{\underset{X}{CH}} \longrightarrow R\,CH_2CH-CH_2-\overset{\cdot}{\underset{X}{CH}}$$

$$\Big\downarrow \begin{array}{c} [CH_2=CH] \\ | \\ X \\ n\text{ steps} \end{array}$$

$$R-\left[CH_2\overset{|}{\underset{X}{CH}}\right]_{n+1}-CH_2-\overset{\cdot}{\underset{X}{CH}}$$

Step (*iii*) **Termination.** The termination of the long chain free radical (formed is step (*ii*)) is achieved by radical coupling or disproportion as given below:

Radical coupling. The radical coupling gives a long chain polymer

$$2\,R\left[CH_2\overset{|}{\underset{X}{CH}}\right]_{n+1}CH_2-\overset{\cdot}{\underset{X}{CH}}$$

$$R\left[CH_2\overset{|}{\underset{X}{CH}}\right]_{n+1}CH_2-\overset{|}{\underset{X}{CH}}-CH-CH_2\left[CH-CH_2\right]_{n+1}R$$

Polypropylene X = H
PVC   X = Cl

Disproportionation

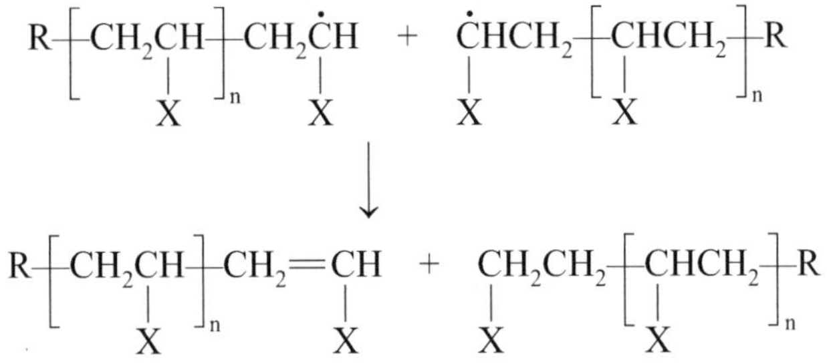

$$R{-}[CH_2CH{-}]_n CH_2\dot{C}H \;+\; \dot{C}HCH_2{-}[CHCH_2{-}]_n R$$
$$\qquad\quad X \qquad\quad X \qquad\quad X \qquad\quad X$$

↓

$$R{-}[CH_2CH{-}]_n CH_2{=}CH \;+\; CH_2CH_2{-}[CHCH_2{-}]_n R$$
$$\qquad\quad X \qquad\quad X \qquad\quad X \qquad\quad X$$

Polyethylene X = H
PVC   X = Cl

As seen, the initator residue (R) is incorporated into the polymer at one or both ends of the chain depending on the type of the termination steps.

2.4.3.4 Carbenes

Carbenes are highly reactive species having a very short half life (< 1 sec.). The parent species, $:CH_2$, is known as methylene. Another example of carbene is dichlorocarbene ($:CCl_2$). As can be seen, carbenes have two nonbonding electrons. These are derived by homolytic fission of two bonds as illustrated below:

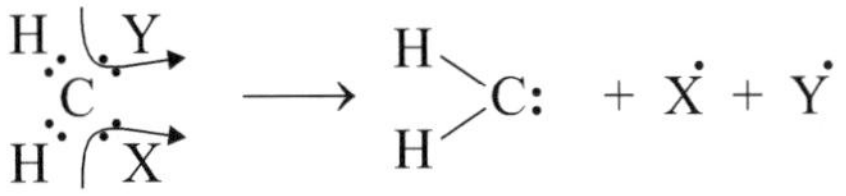

The carbon atom in a carbene has six electrons and so it is an electron deficient species analogous to carbocation. Thus, carbene normally reacts as a strong electrophilie.

A carbene is known to exist as *singlet* or *triplet* carbene. In case, both the electrons occupy the same orbital (*i.e.*, have antiparallel spins), the carbene exists in singlet state. There is no magnetic moment in this state. On the other hand, if the two electrons occupy different orbitals, they will have parallel (unpaired) spins; and such a carbene exists in triplet state and would have magnetic moment.

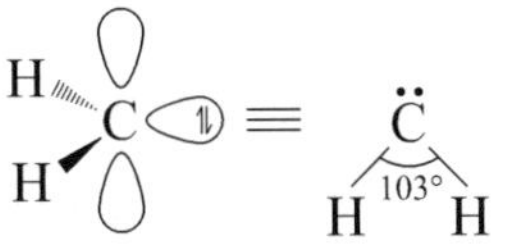

Singlet state of carbene 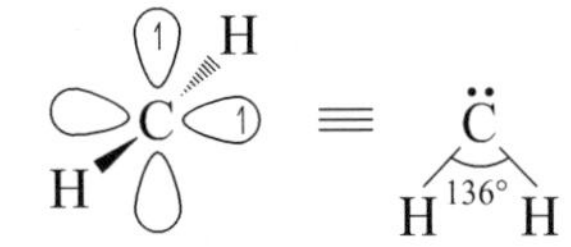   Triplet state of carbene

In the singlet state of carbene (or methylene), the carbon atom is sp^2 hybridised. Two of the sp^2 orbitals are bonded to the hydrogen atoms and the third contains a lone pair of electrons. The unhybridised *p* orbital is unoccupied. The HCH bond angle is 103° and the C—H bond length is 1.12 Å.

On the other hand, in the triplet state of carbene, the carbon atom is *sp* hybridised. The two *sp* hybridised orbitals are bonded to two hydrogen atoms and the two unhybridised p orbitals contain one electrons each. The triplet state of carbene is also a bent molecule with an angle of about 136° and C—H bond length of 103 Å.

Whereas a singlet carbene is *diamagnetic*, the triplet carbene is *paramagnetic*.

Carbenes in which the carbene carbon is attached to two atoms, each bearing a lone pair of electrons, are more stable due to resonance.

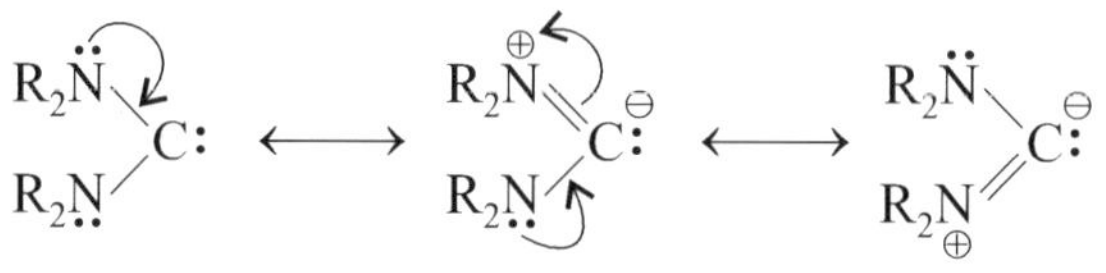

$$R_2\ddot{N}\diagdown \atop R_2\ddot{N}\diagup C: \quad\longleftrightarrow\quad R_2\overset{\oplus}{N}\diagdown \atop R_2\ddot{N}\diagup \overset{\ominus}{C}: \quad\longleftrightarrow\quad R_2\ddot{N}\diagdown \atop R_2\underset{\oplus}{N}\diagup \overset{\ominus}{C}:$$

Generation of Carbenes

(*i*) **From alkyl halides.** This is the most common method used for the generation of carbenes. Thus, an alkyl halide on treatment with base generate dichlorocarbene.

$$CHCl_3 \xrightarrow{\text{Base}} {}^-CCl_3 \longrightarrow :CCl_2 + Cl^-$$

Chloroform   Dichloro
                                           Carbene

In the above reaction, a strong base like $KOC(CH_3)_3$ is used and the reaction is carried out in anhydrous solvent since in aqueous solution the dichloro carbene may undergo hydrolysis. The reaction takes place in two steps as shown below:

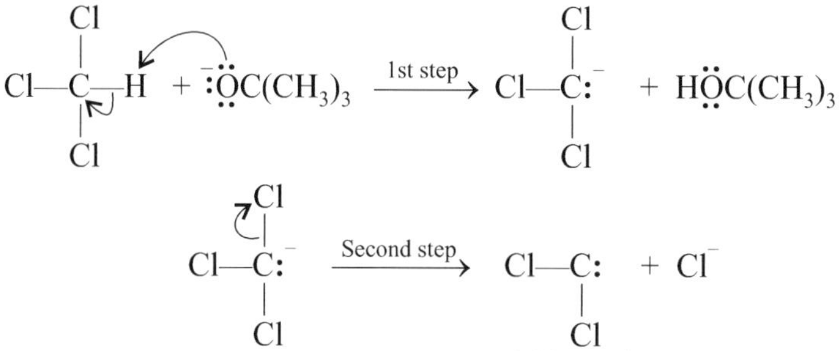

$$Cl-\underset{\underset{Cl}{|}}{\overset{\overset{Cl}{|}}{C}}-H + {}^-\ddot{O}C(CH_3)_3 \xrightarrow{\text{1st step}} Cl-\underset{\underset{Cl}{|}}{\overset{\overset{Cl}{|}}{C}}: + H\ddot{O}C(CH_3)_3$$

$$Cl-\underset{\underset{Cl}{|}}{\overset{\overset{Cl}{|}}{C}}:^- \xrightarrow{\text{Second step}} Cl-\underset{\underset{Cl}{|}}{C}: + Cl^-$$

Dichlorocarbene

In the above reaction using aqueous sodium hydroxide, the yield is less than 5 percent. It has been found that using a phase transfer catalyst (like benzyl triethylammonium chloride) the yield improves to about 60-70 percent.

$$CHCl_3 + aqKOH \xrightarrow[\text{60-70\%}]{\text{PTC}} :CCl_2$$

The reason for better yield using PTC is that as formed dichlorocarbene is transferred to the organic phase (as soon as it is generated)

(*ii*) **From Ketenes.** Carbenes are generated by thermal or photolytic decomposition of ketenes. The required ketene is obtained from diazoketones by pyrolysis.

$$\begin{matrix} H_3C \\ \quad\quad C=O \\ H_3C \end{matrix} \xrightarrow{\Delta} \underset{\text{ketene}}{CH_2=C=O}$$

$$\underset{}{R-\overset{\overset{\displaystyle O}{\|}}{C}-Cl} \xrightarrow{CH_2N_2} \underset{\text{Diazoketone}}{R-\overset{\overset{\displaystyle O}{\|}}{C}-NH_2} \xrightarrow{Ag_2O} \underset{\text{Ketene}}{R\ CH=C=O:} + N_2$$

$$\begin{matrix} R \\ \quad\quad C=C=O: \\ R \end{matrix} \underset{\text{ketene}}{\quad} \xrightarrow[\text{or }\Delta]{h\nu} \begin{matrix} R \\ \quad\quad C: \\ R \end{matrix} + CO$$
$$\underset{\text{carbene}}{\quad}$$

(*iii*) **From epoxides.** Epoxides (prepared from alkenes by oxidation with per acids) on photolytic decomposition generate carbenes.

$$\underset{\text{Alkene}}{\begin{matrix} R \quad\quad R \\ C=C \\ R \quad\quad R \end{matrix}} \xrightarrow{RCO_3H} \underset{\text{Epoxide}}{\begin{matrix} R \quad O \quad R \\ C-C \\ R \quad\quad R \end{matrix}} \xrightarrow{h\nu} \underset{\text{Carbene}}{\begin{matrix} R \\ C: \\ R \end{matrix}} + \underset{\text{Ketone}}{\begin{matrix} O \\ \| \\ C \\ R \quad R \end{matrix}}$$

(*iv*) **From diazirines.** Carbenes are generated by the decomposition of diazirines.

$$\underset{\text{Diazirines}}{\begin{matrix} R \quad N \\ C \ \| \\ R \quad N \end{matrix}} \xrightarrow[\text{or }\Delta]{h\nu} \underset{\text{Carbene}}{\begin{matrix} R \\ C: \\ R \end{matrix}} + N_2$$

The required diazirines are prepared by the reaction of ammonia or primary amine with ketones in presence of chloramine. The formed diazirides on treatments with Ag$_2$O give the required diazirines.

$$\begin{matrix} R \\ \quad C=O \\ R \end{matrix} + \begin{matrix} NH_3 \\ \text{or} \\ RNH_2 \end{matrix} \xrightarrow{:NH_2Cl} \begin{matrix} R \\ \quad C=N-H \\ R \end{matrix} + :NH_2Cl$$

$$\downarrow H$$

$$\underset{\text{Diazirines}}{\begin{matrix} R \quad N \\ C \ \| \\ R \quad N \end{matrix}} \xleftarrow{Ag_2O} \underset{\text{Diaziridines}}{\begin{matrix} R \quad NH \\ C \ | \\ R \quad NH \end{matrix}} \xleftarrow[HCl]{Base} \begin{matrix} R \quad N-H \\ C \\ R \quad N-Cl \\ \quad\quad H \end{matrix}$$

(v) **From aliphatic diazo compounds.** Aliphatic diazo compounds on decomposition either photochemically or thermally generate carbene.

$$\underset{R}{\overset{R}{\diagdown}}C{=}N_2 \xrightarrow{\ h\nu\ } \underset{R}{\overset{R}{\diagdown}}C: \ + \ N_2$$

$$H_2C{=}\overset{\ominus}{N}{=}\overset{\oplus}{N}: \ \longleftrightarrow \ :\overset{-}{C}H_2{-}\overset{+}{C}{\equiv}N \xrightarrow[-N_2]{\ h\nu\ } \underset{H}{\overset{H}{\diagdown}}C:$$

Diazomethane $\qquad\qquad\qquad\qquad\qquad$ Carbene

$$R{-}COCHN_2 \xrightarrow{\ h\nu\ } RCOCH: + N_2$$

Acyl diazo compound $\qquad$ Acyl carbene

$$N_2CHCOOC_2H_5 \xrightarrow{\ h\nu\ } :CHCOOC_2H_5 + N_2$$

Diazoacetic ester $\qquad\qquad$ Carbethoxy carbene

Reactions of Carbenes

Following are given some of the reactions of carbenes

(i) **Addition reactions.** Carbenes add on to C=C to give cyclopropane derivatives. The required carbene is generated from chloroform and aq. NaOH using PTC. Some addition reactions of carbenes are given below:

$$\text{Cyclohexene} + CHCl_3 \xrightarrow[\text{PTC}]{\text{aq. NaOH}} \text{(bicyclic product)} \underset{\text{60–70\%}}{\overset{Cl}{\underset{Cl}{}}}$$

$$\underset{\text{Styrene}}{C_6H_5CH{=}CH_2} \xrightarrow[\text{PTC}]{\text{aq. NaOH}} C_6H_5CH{-}CH_2$$

$$\underset{\substack{Cl\quad Cl \\ \text{1-Phenyl-2,2-dichloro} \\ \text{cyclopropane}}}{}$$

Carbenes also add to C=N bond of Schiffs base

$$C_6H_5{-}CH{=}N{-}C_6H_5 + CHCl_3 \xrightarrow[C_6H_5CH_2\overset{+}{N}Et_3Cl^-]{\text{aq. NaOH}} C_6H_5{-}CH{-}N{-}C_6H_5$$

$$\overset{Cl\quad Cl}{}$$

(ii) **Insersion into C—H bonds.** Carbenes can insert into C—H single bonds

$$-\overset{|}{\underset{|}{C}}{-}H \ : \ CH_2 \ \longrightarrow \ -\overset{|}{\underset{|}{C}}{-}CH_2{-}H$$

The reaction is believed to occur as follows:

$$-\overset{|}{\underset{|}{C}}{-}H + :CH_2 \longrightarrow \left[-\overset{|}{\underset{|}{C}}\underset{CH_2}{\overset{-H}{\cdots}} \right] \longrightarrow -\overset{|}{\underset{|}{C}}{-}CH_2{-}H$$

A typical insertion reaction is given below:

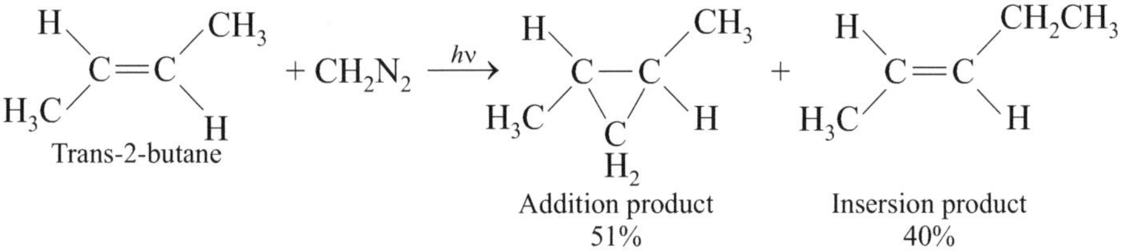

Trans-2-butane

Addition product
51%

Insertion product
40%

Some other insertion reactions are given below:

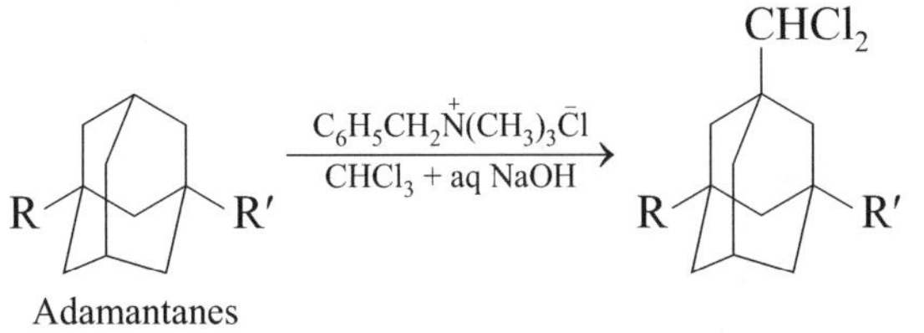

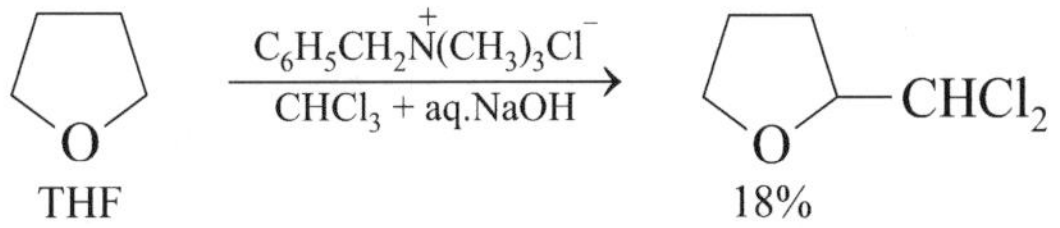

Adamantanes

THF

18%

(*iii*) **Reaction with hydrazine: Formation of Diazomethane.** The reaction of hydrazine with dichlorocarbene (generated by the PTC method) gives diazomethane

$$NH_2NH_2 + CHCl_3 + NaOH \xrightarrow[\substack{Ether \\ or \\ CH_2Cl_2}]{PTC} CH_2N_2 \text{ in ether or } CH_2Cl_2$$

Hydrazine   35%

In this case use of crown ether in place it PTC gives better yields (48%).

(*iv*) **Reaction with primary amines (synthesis of isonitriles).** Dichlorocarbene (generated by PTC method) reacts with primary amines to yield isonitriles

$$RNH_2 + CHCl_3 + NaOH \xrightarrow[aq]{C_6H_5CH_2\overset{+}{N}Et_3Cl^-} R{-}N{\equiv}C$$

40—60 %
isonitriles

The above reaction is called **Phase Transfer Hofmann Carbylamine Reaction**.

(*v*) **Synthesis of nitriles.** Dichlorocarbene (generated by the PTC method) on reaction with amides, thioamides, aldoximes and amidines give the corresponding nitriles

$$\left.\begin{array}{l} RCONH_2 \\ RCSNH_2 \\ RCH{=}NOH \\ RC{-}NH_2 \\ \;\;\|\!\! \\ \;\;NH \end{array}\right] + CHCl + NaOH \xrightarrow[aq.]{C_6H_5CH_2\overset{+}{N}Et_3Cl^-} R{-}CN$$

The formation of nitriles from amides is shown below:

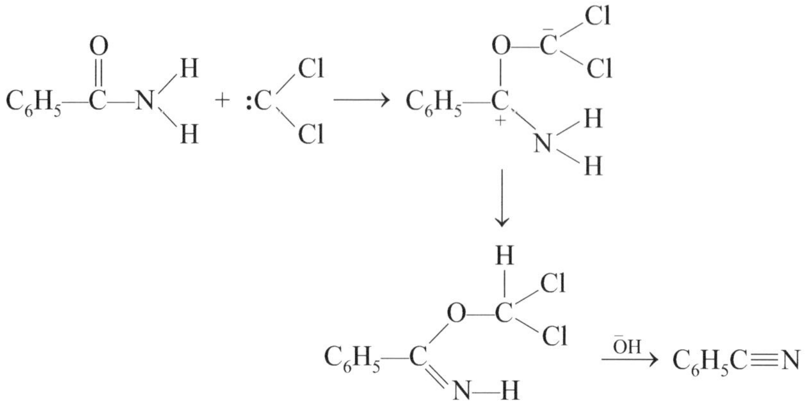

(*vi*) **Reaction with Alcohols.** Reaction of alcohols with dichloro carbene give good yield of chlorides

$$ROH + CHCl_3 + NaOH \xrightarrow[ar]{C_6H_5CH_2\overset{+}{N}Et_3Cl^-} RCl + NaCl + H_2O$$

In case of steroidal alcohols the OH is replaced with Cl with relention of configuration.

(*vii*) **Reaction with aldehydes: synthesis of mandelic acids.** Dichlorocarbene react with aromatic aldehydes to give mandelic acids.

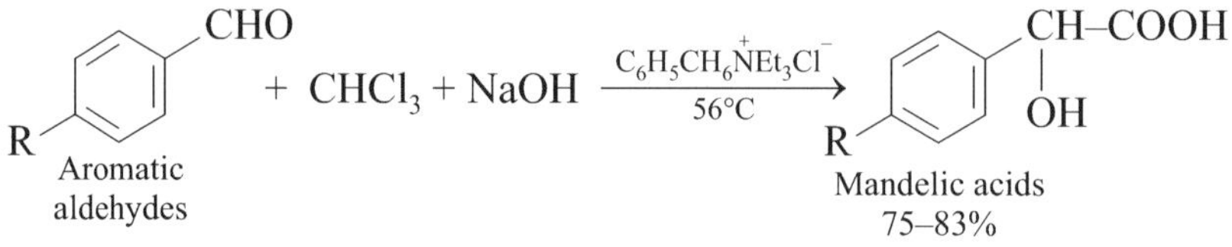

The above reaction proceeds via the addition of :CCl$_2$ to carbonyl group

(of aldehydes) to give R—CH—O, which on alkaline hydrolysis give

the desired product.

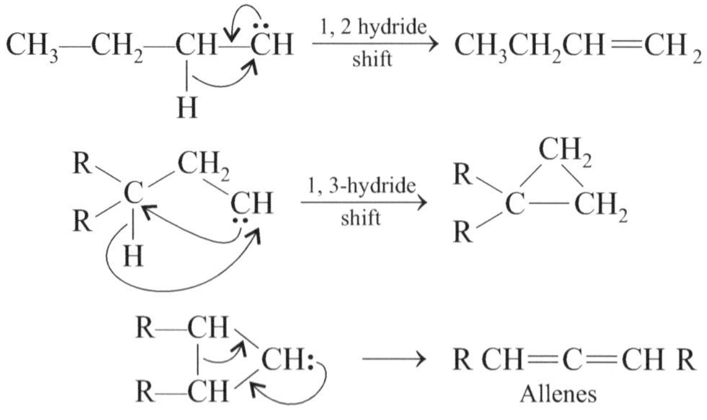

(*viii*) **Rearrangements.** Alkyl carbenes undergo rearrangements involving migration of alkyl group or hydrogen

(*ix*) **Ring expansion.** In some cases, addition of carbenes involve ring expansion. As an example, the reaction of indene with dichlorocarbene gives 2-chloronaphthalene

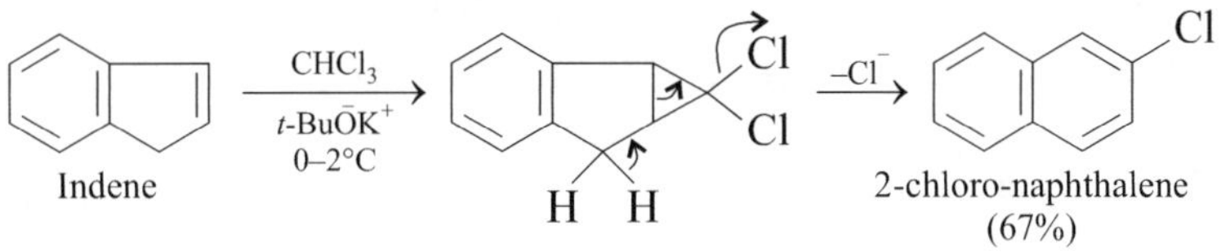

Indene
2-chloro-naphthalene
(67%)

Some other examples of ring expansion are given below:

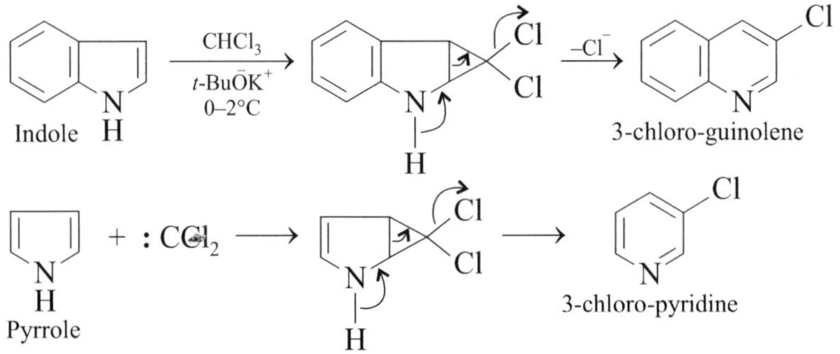

Indole   H
3-chloro-guinolene

Pyrrole
3-chloro-pyridine

(*x*) **Ring contraction.** The decomposition of cyclic diazoketones give the corresponding carbene, which undergoes ring contraction

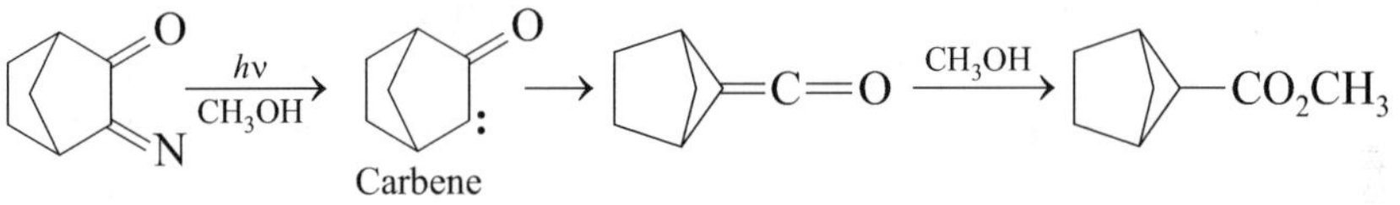

Carbene

(*xi*) **Synthesis of allenes.** Dihalocyclopropanes on heating with sodium or magnesium in ether give allenes.

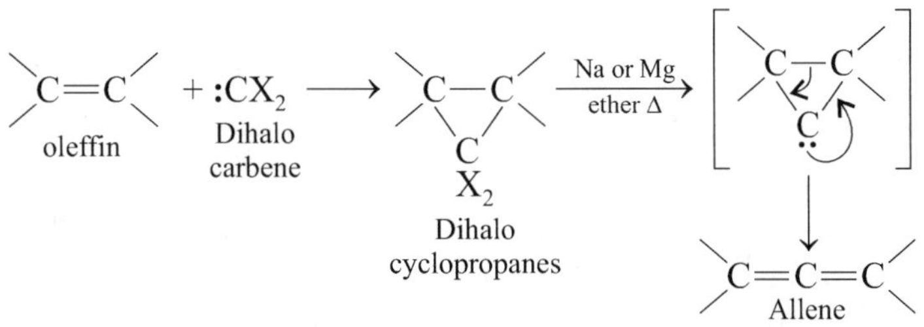

oleffin   Dihalo carbene
Dihalo cyclopropanes
Allene

(*xii*) **Synthesis of spiro comounds.** Reaction of cyclohexene with :CBr$_2$ followed by treatment of the adduct will CH$_3$Li and cyclohexene gives spiro compounds

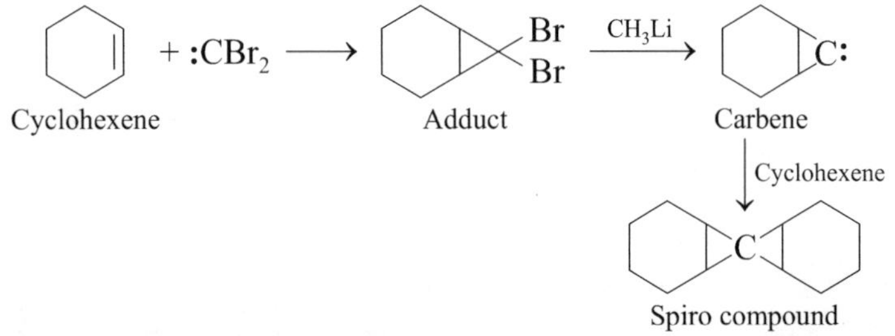

Cyclohexene
Adduct
Carbene
Cyclohexene
Spiro compound

Reactions Involving Carbenes

Carbenes are obtained in a number of reactions as intermediates. These include carbylamine reaction, Wolff rearrangement and Reimer Tiemann reaction.

(*i*) **Carbylamine reaction.** The formation of isonitriles by the reaction of primary amines with chloroform in the presence of alkali is known as *carbylamine reaction.* This reaction is used as a test for primary amine.

$$RNH_2 + CHCl_3 + 3NaOH \longrightarrow \quad RN{\equiv}C \quad + 3NaCl + 3H_2O$$
Isonitrile
(has offensive odour)

The reaction occurs via the formation of dichlorocarbene (:CCl_2) intermediate, generated in situ.

$$CHCl_3 + \bar{O}H \longrightarrow \bar{C}Cl_3 + H_2O$$

$$\bar{C}Cl_3 \longrightarrow \quad :CCl_2 \quad + Cl^-$$
Dichlorocarbene

$$R\ddot{N}H_2 + :CCl_2 \longrightarrow RNHCHCl_2$$

$$RNHCHCl_2 + 2\bar{O}H \longrightarrow R{-}\overset{+}{N}{\equiv}\bar{C} + 2Cl^- + 2H_2O$$
Isonitrile

(*ii*) **Wolff rearrangement.** α-Diazoketones (prepared by the reaction of an acid chloride with diazo compound like diazomethane) form ketenes in the presence of silver oxide. This reaction is known as Wolff rearrangement.

$$\underset{\text{Acid chloride}}{R{-}\overset{\overset{\displaystyle O}{\|}}{C}{-}Cl} + 2CH_2N_2 \xrightarrow{Ag_2O} \underset{\text{α-Diazoketone}}{R{-}\overset{\overset{\displaystyle O}{\|}}{C}{-}\bar{C}H{-}\overset{+}{N}{\equiv}N} + CH_3Cl + N_2$$

$$R{-}\overset{\overset{\displaystyle O}{\|}}{C}{-}\overset{+}{C}H{-}\overset{+}{N}{\equiv}N \xrightarrow[-N_2]{Ag_2O} \underset{\text{Ketene}}{R{-}CH{=}C{=}O}$$

It may be noted here that Wolff rearrangement generates ketene in the absence of any nucleophile and, therefore, the product can be isolated from the reaction mixture. However, when this rearrangement is carried out in the presence of water, alcohol or amine, the ketene is converted into carboxylic acid, ester or amide respectively. This reaction is called **Arndt-Eistert synthesis**.

$$\underset{\text{Ketene}}{R{-}\overset{\overset{\displaystyle H}{|}}{C}{=}C{=}O} \quad
\begin{array}{l}
\xrightarrow{R'OH} RCH_2COOR' \text{ (Ester)} \\
\xrightarrow{H_2O} RCH_2COOH \text{ (Carboxylic acid)} \\
\xrightarrow{NH_3} RCH_2CONH_2 \text{ (Amide)}
\end{array}$$

The mechanism of Wolff is explained below:

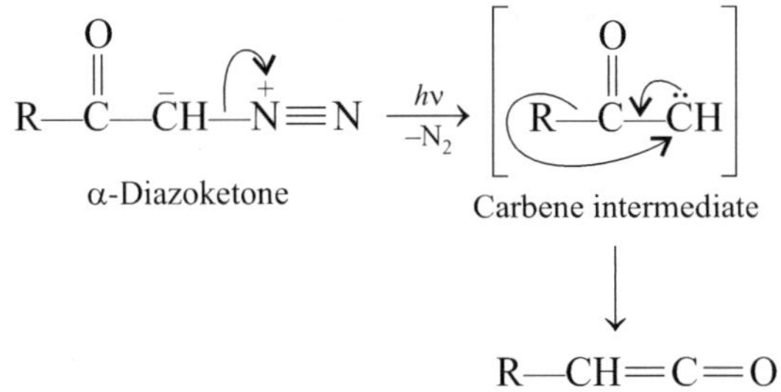

α-Diazoketone   Carbene intermediate

$$R—CH=C=O$$
Ketene

(*iii*) **Reimer-Tiemann reaction.** Phenols on reaction with cloroform in presence of sodium hydroxide (or potassium hydroxide) solution afford hydroxy aldehydes. The formyl group is introduced at *ortho* position unless one or both the ortho positions are occupied; in that case, the attack is directed at *para* position. This formylation reaction is known as Reimer-Tiemann reaction.

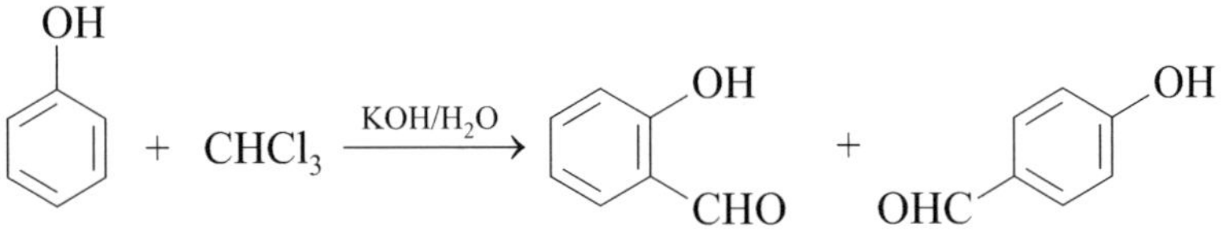

Phenol   Salicylaldehyde   4-Hydroxybenzaldehyde
                          (major product)   (minor product)

The reaction proceeds via the formation of dichlorocarbene, and the various steps involved are shown below:

$$CHCl_3 \xrightarrow{\bar{O}H} \bar{C}Cl_3 + H_2O$$

$$\bar{C}Cl_3 \xrightarrow{-Cl} \quad :CCl_2$$
Dichlorocarbene

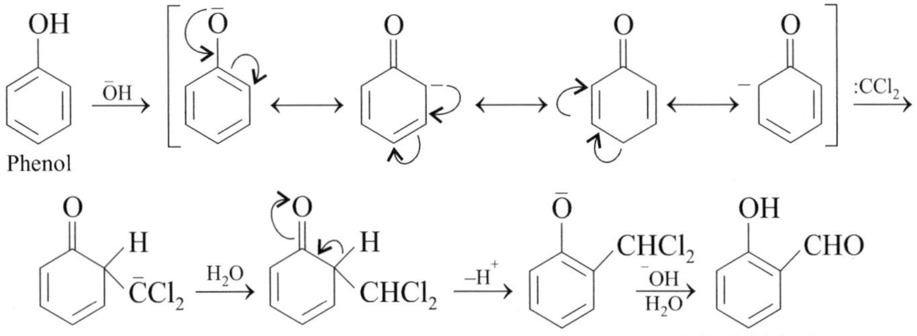

Phenol

Salicylaldehyde

2.4.3.5 Nitrenes

Nitrenes are the   nitrogen analogues of carbenes and are electron deficient monovalent nitrogen species. As in the case of carbenes, the nitrenes too exist in *singlet* and *triplet* state.

$$R - \ddot{N}:↿⇂ \qquad R - \ddot{N}:↿↿$$

Singlet nitrene   Triplet nitrene

Nitrenes are very reactive species and cannot be isolated. However a nitrene can be trapped by reacting with carbon monoxide and ethylene.

$$PhN_3 \xrightarrow[-N_2]{\Delta} Ph\ddot{N}: \xrightarrow{CO} PhN{=}C{=}O$$

Phenyl isocyanate

$$H\ddot{N}: + CH_2{=}CH_2 \rightleftharpoons \underset{H}{\overset{\triangledown}{N}}$$

Generation of Nitrenes

(*i*) **From azides.** Nitrenes can be conveniently generated by thermal or photolytic decomposition of azides or acyl azides.

$$R{-}\bar{N}{-}\overset{+}{N}{\equiv}N \xrightarrow{h\nu} R{-}\ddot{N} + N_2$$

Azide   Nitrene

$$RCON_3 \xrightarrow{h\nu} RCO\ddot{N} + N_2$$

Acyl azide   Acyl nitrene

(*ii*) **From Sulfinylamines.** Photolysis of sulfinylamines generate nitrene

$$Ph{-}N{=}S{=}O \xrightarrow[\text{gas phase}]{\Delta} Ph{-}\ddot{N} + SO$$

Phenyl Sulfinylamine   Nitrene

(*iii*) **From N-benzenesulfonoxy carbamates.** Nitrenes are geneated by treatment of N-benzenesulfonoxy carbamates with a base.

$$\underset{\substack{\text{N-Benzene-sulfonoxy}\\\text{carbamate}}}{H_5C_2O\overset{\overset{\displaystyle O}{\|}}{C}NHOSO_2-\bigcirc} \xrightarrow{{}^-OH} \underset{\substack{\text{Carbethoxy}\\\text{nitrene}}}{H_5C_2O\overset{\overset{\displaystyle O}{\|}}{C}-\ddot{N}} + C_6H_5SO_3^-$$

Reactions of Nitrenes

As already mentioned addition of nitrenes to C=C yields aziridines.

$$\underset{/}{\overset{\backslash}{C}}{=}\underset{\backslash}{\overset{/}{C}} + R\ddot{N} \longrightarrow \underset{\text{Aziridines}}{\underset{/}{\overset{\backslash}{C}}{-}\underset{\backslash}{\overset{/}{C}}}$$
(with N bridging, R—N)

(*i*) **Insersion.** Nitrenes particularly acyl nitrenes can insert into C—H bonds. A typical example is given below.

$$R-\underset{\underset{O}{\|}}{C}-\ddot{\underset{..}{N}} \quad + R_3C-H \longrightarrow R_3C-NH-\underset{\underset{O}{\|}}{C}-R'$$

Acylnitrine

A useful reaction involves annulation of pyrroles and thiophenes

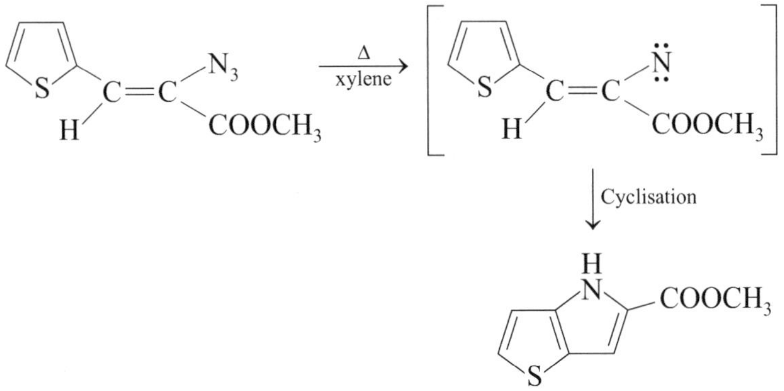

Cyclisation

(*ii*) **Dimerisation.** Dimerisation of arylnitrene give azobenzene

$$2Ar\ddot{\underset{..}{N}} \longrightarrow ArN{=}NAr$$

Arylnitrene

(*iii*) **Abstraction of Hydrogen.** Abstraction of hydrogen from carbon alpha to the nitrogen leads to the formation of imines. However, abstraction of hydrogen takes place from position 4, followed by ring closure give pyrrolidines

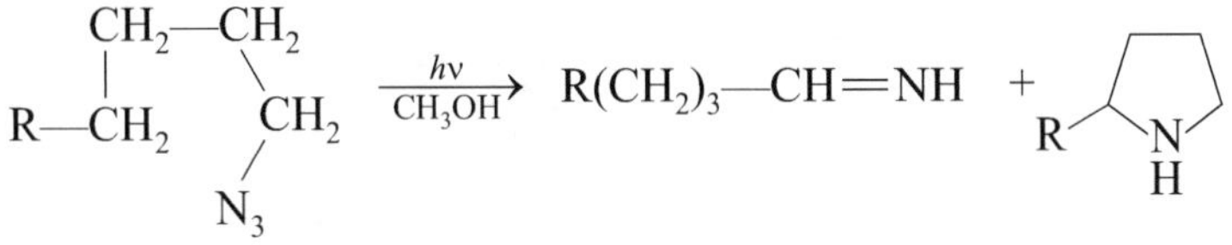

Reactions Involving Nitrenes

Nitrenes are formed as intermediates in a number of reactions like Hofmann, Curtius, Schmidt and Lossen rearrangements.

(*i*) **Hofmann rearrangement.** In this rearrangement reaction, a primary amide is converted into an amine (having one carbon less than the starting amide) by the reaction of sodium hypohalite (usually generated in situ from halogen and sodium hydroxide).

$$CH_3CH_2CONH_2 + Br_2 + 4KOH \longrightarrow$$
Propanamide

$$CH_3CH_2NH_2 + 2KBr + K_2CO_3 + 2H_2O$$
Ethylamine

The reaction occurs via the formation of nitrene intermediate as explained below:

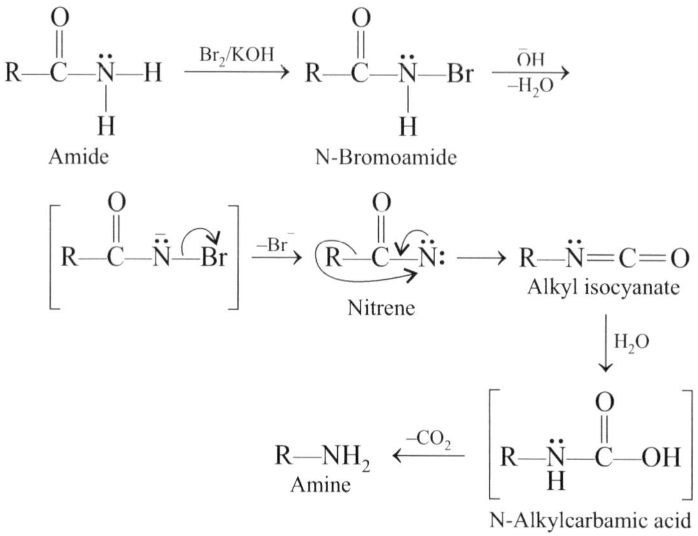

Amide → N-Bromoamide

Nitrene → R—N̈=C=O Alkyl isocyanate

R—NH$_2$ (Amine) $\xleftarrow{-CO_2}$ N-Alkylcarbamic acid

(*ii*) **Curtius rearrangement.** Rearrangement of acid azides under the influence of heat or light in a non-aqueous solvent such as chloroform, benzene or ether to an isocyanate is known as *Curtius rearrangement*. The isocyanate on hydrolysis then gives a primary amine. Thus, this reaction serves to convert an acid azide into an amine with one carbon atom less than the azide.

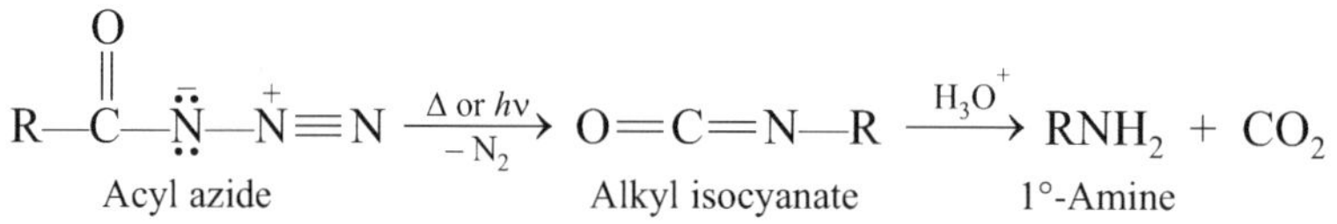

$$R-\overset{O}{\overset{||}{C}}-\overset{..}{\underset{..}{N}}-\overset{+}{N}\equiv N \xrightarrow[-N_2]{\Delta \text{ or } h\nu} O=C=N-R \xrightarrow{H_3O^+} RNH_2 + CO_2$$

Acyl azide   Alkyl isocyanate   1°-Amine

The starting azides are usually prepared by the action of sodium azide on acid chlorides or by the action of nitrous acid upon acid hydrazides.

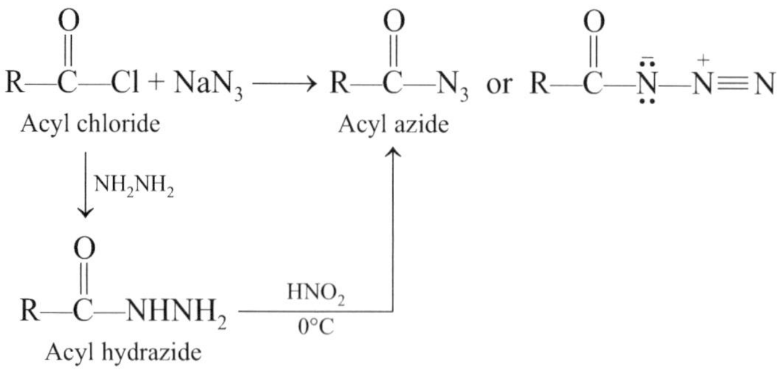

$$R-\overset{O}{\overset{||}{C}}-Cl + NaN_3 \longrightarrow R-\overset{O}{\overset{||}{C}}-N_3 \text{ or } R-\overset{O}{\overset{||}{C}}-\overset{..}{\underset{..}{N}}-\overset{+}{N}\equiv N$$

Acyl chloride   Acyl azide

R—C(=O)—NHNH$_2$ (Acyl hydrazide) $\xrightarrow[0°C]{HNO_2}$

Mechanism of the rearrangement:

$$R-\overset{O}{\overset{||}{C}}-\ddot{\overset{+}{N}}-N\equiv N \xrightarrow[-N_2]{\Delta} O=C=N-R$$

Acid azide   Alkyl isocyanate

$$R-N=C=O \xrightarrow{H\ddot{O}H} \left[R-\overset{..}{N}-\overset{\overset{O}{||}}{C}-OH \atop \underset{H}{|} \right] \xrightarrow{-CO_2} RNH_2$$

Alkyl isocyanate   1°-Amine

The existence of nitrene intermediate is supported by the following sequence of reactions:

$$R_3C-\overset{\overline{\ddot{}}}{\underset{..}{N}}-\overset{+}{N}-N\equiv N \xrightarrow[-N_2]{\Delta} \overset{R}{\underset{R}{\overset{|}{R-C-\overset{..}{\underset{..}{N}}}}} \longrightarrow \overset{R}{\underset{R}{\diagup}}C=N-R$$

tert. alkyl azide   Nitrene

(*iii*) **Lossen arrangement.** The rearrangement of acyl derivatives of hydroxamic acid into isocyanate in presence of a strong inorganic acid, followed by hydrolysis to give the corresponding amine is known as *Lossen rearrangement.*

$$R-\overset{\overset{O}{||}}{C}-NH-OH \longrightarrow R-\overset{\overset{O}{||}}{C}-NH-O-\overset{\overset{O}{||}}{C}-R'$$

Hydoxamic acid   Acyl derivative of hydroxamic acid

$$\Big\downarrow \overline{O}H$$

$$O=C=N-R \xleftarrow{-R'COO^-} \left[R-\overset{\overset{O}{||}}{C}-\overset{..}{N}-O-\overset{\overset{O}{||}}{C}-R' \right.$$

Alkyl isocyanate

$$\Big\downarrow H_2O$$

$$RNH_2 + CO_2$$

Amine

$$\left. R-C=N-O-\overset{\overset{O}{||}}{C}-R' \right]$$

Hydroxamic acid itself may also undergo lossen rearrangement in presence of strong inorganic acids to give a primary amine.

$$R-\overset{\overset{O}{||}}{C}-\overset{\overset{H}{|}}{N}-OH \xrightarrow{HCl} R-\overset{\overset{O}{||}}{C}-\overset{\overset{H}{|}}{\underset{..}{N}}-\overset{+}{O}H_2 \xrightarrow{-H_2O} R-\overset{\overset{O}{||}}{C}-\overset{\overset{H}{|}}{\underset{..}{\overset{+}{N}}}$$

Hydroxamic acid

$$\xrightarrow{-H^+} \left[R-\overset{\overset{O}{||}}{C}-\overset{..}{N} \right] \longrightarrow O=C=N-R \xrightarrow{-H_2O} RNH_2 + CO_2$$

Nitrene   Isocyanate   1°-Amine

2.4.3.6 Benzyne

Benzynes are neutral, highly reactive reaction intermediates in which the aromatic character is not markedly disturbed. Benzynes (or arynes) contain a carbon–carbon triple bond and may be regarded as the aromatic counterpart of acetylene. In fact, benzyne is benzene minus two o-hydrogens and so it is called dehydrobenzene. It is believed that the new bond of benzyne is formed by the overlap of sp^2 orbitals belonging to two neighbouring carbon atoms. As the overlapping on the sides is not very effective, the new bond is weak and so benzyne is a highly reactive species.

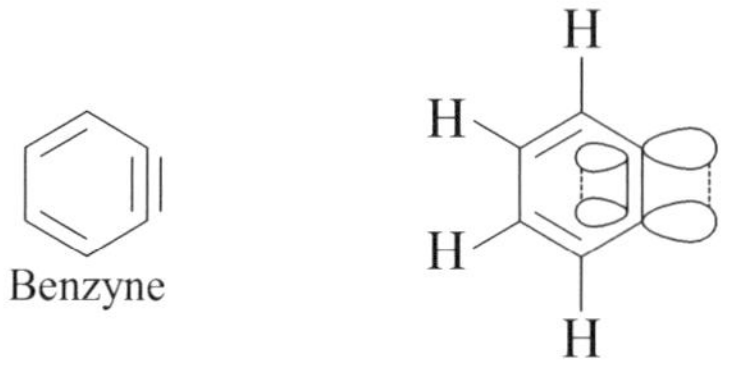

Benzyne

Generation of Benzynes

(i) From aryl halides. Benzene can be generated from aryl halides on treatment with a strong base like KNH_2, PhLi, etc.

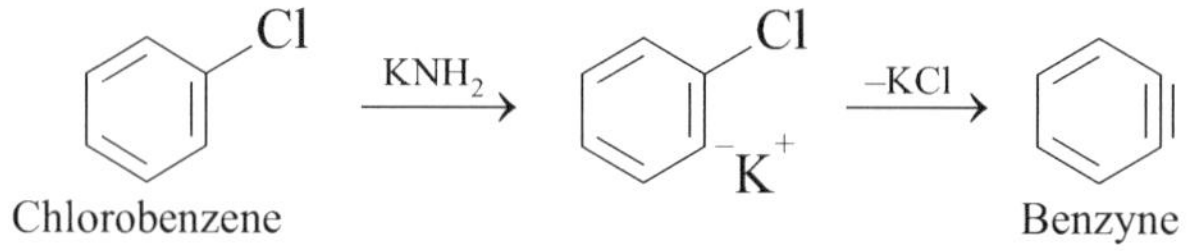

Chlorobenzene   Benzyne

(ii) From o-Aminobenzoic acid. Benzyne can also be generated from o-aminobenzoic acid by diazotisation followed by decomposition of the resulting diazo compound.

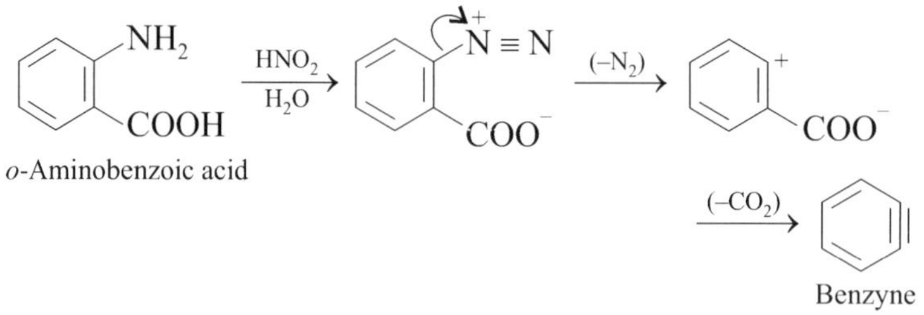

o-Aminobenzoic acid

Benzyne

(iii) From phthaloyl Peroxide. Phthaloyl peroxide on photolytic decomposition generates benzyne via the formation of a lactone intermediate.

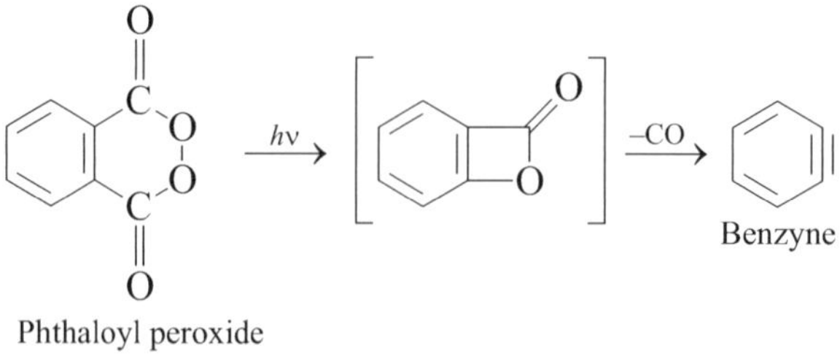

Phthaloyl peroxide

(*iv*) From Benzothiazoate 1, 1-dioxide. Benzothiazoate 1, 1-dioxide on photolytic decomposition generates benzyne

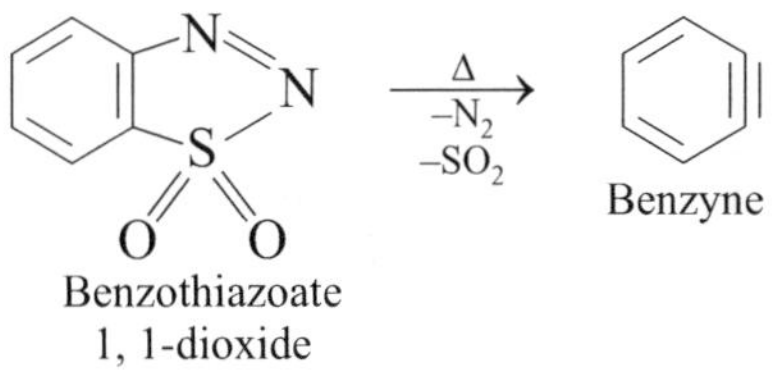

Benzothiazoate
1, 1-dioxide

Benzyne

(*v*) From benzenetrifluoromethane sulfonate. Benzenetrifluoromethane sulfonate on treatment with a strong base generates benzyne

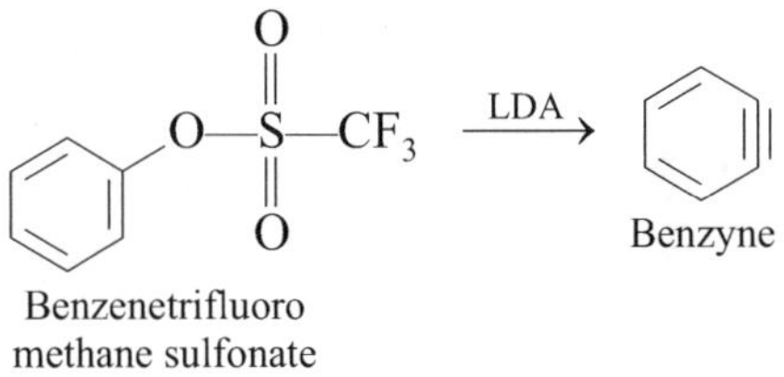

Benzenetrifluoro
methane sulfonate

Benzyne

Reactions of Benzynes

(*i*) Reaction with nucleophiles. It has already been stated that aryl halides on treatment with KNH_2 (a strong base) generate benzyne. Further, when KNH_2 is used, the available nucleophile ($\overline{N}H_2$) reacts with the generated benzyne to give aniline.

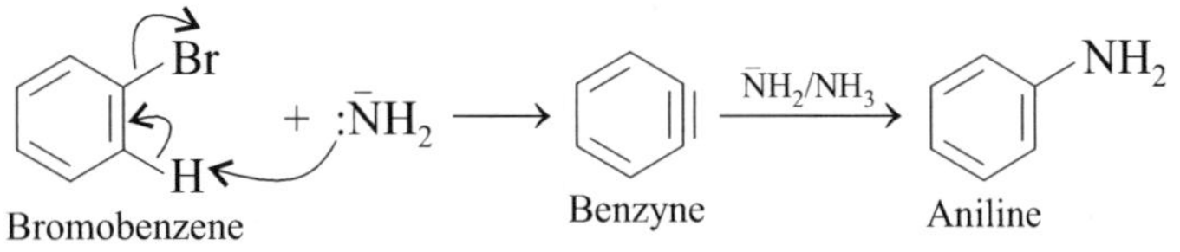

Bromobenzene

Benzyne

Aniline

Using *p*-chlorotoluene in the above reaction (in place of bromobenzene), a mixture of para and meta toluidines is obtained

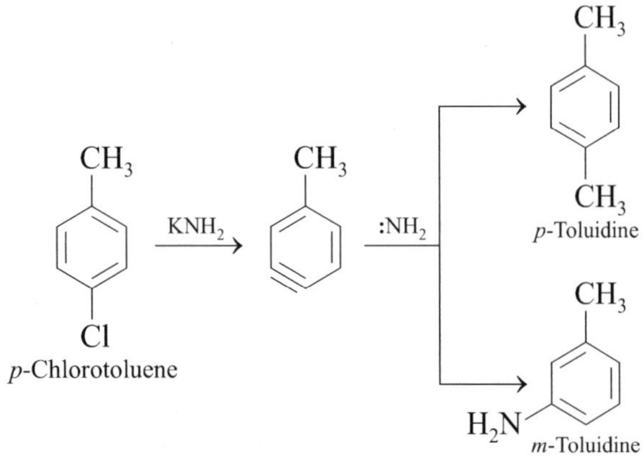

p-Chlorotoluene

p-Toluidine

m-Toluidine

However, treatment of both *o*-bromoanisole and *m*-bromoanisole with KNH_2/NH_3 gives *m*-aminoanisole in 95 percent yield.

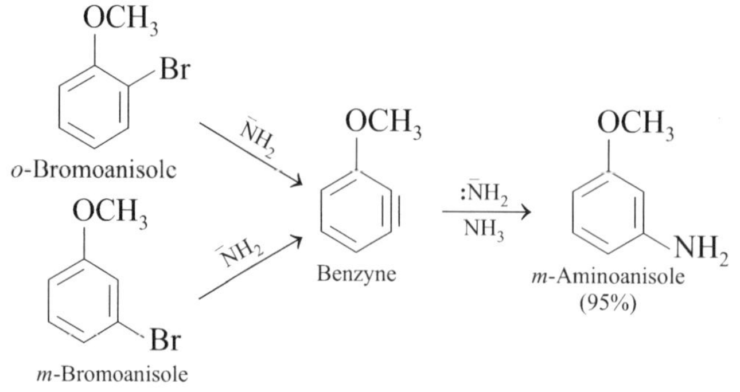

o-Bromoanisole

m-Bromoanisole

Benzyne

m-Aminoanisole
(95%)

(*ii*) Synthesis of t-butyloxyphenol, *o*-phenyl cyclohexanone and phenyl benzoate

The title compounds are synthesised as given below:

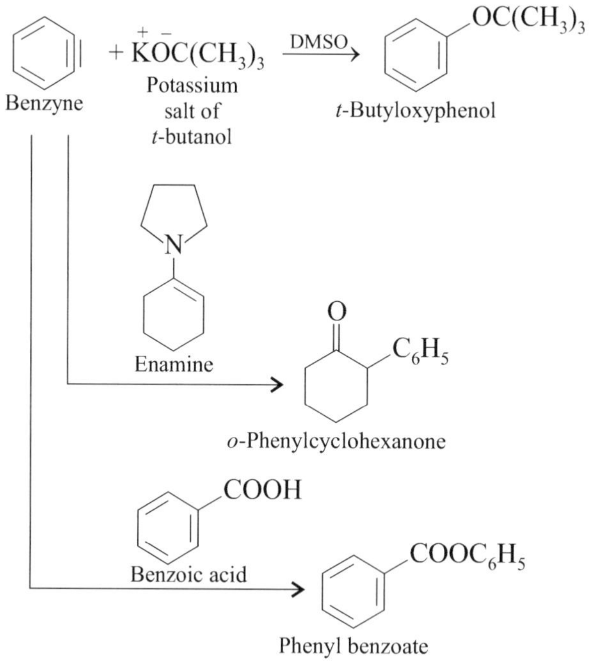

Benzyne

Potassium salt of t-butanol

t-Butyloxyphenol

Enamine

o-Phenylcyclohexanone

Benzoic acid

Phenyl benzoate

(*iii*) **Reaction with anthracene, furan and aniline.** Benzyne reacts with anthracene, furan and aniline to give triptycene, α-naphthol and N-substituted anilines respectively.

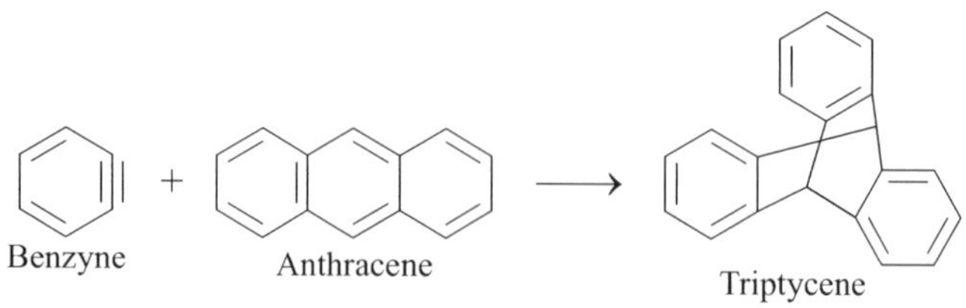

Benzyne   Anthracene

Triptycene

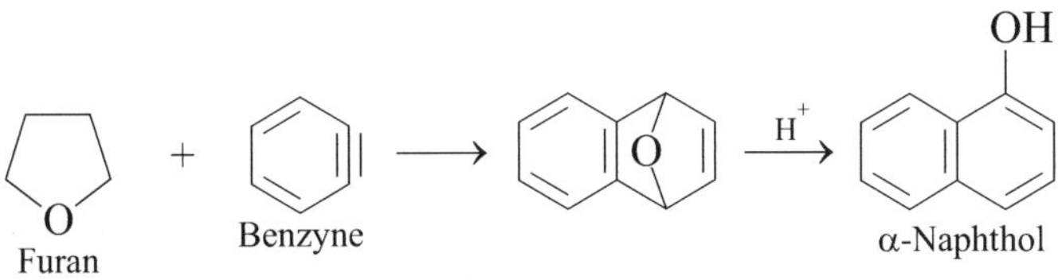

Furan + Benzyne ⟶ ⟶ α-Naphthol

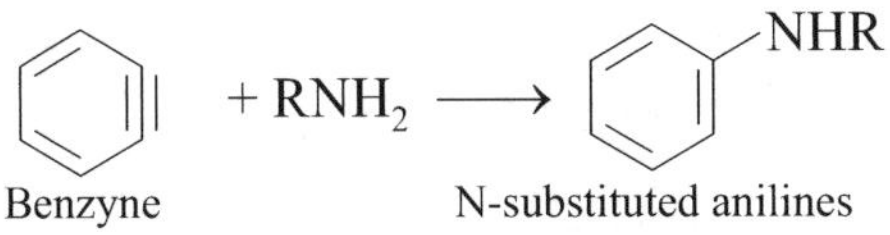

Benzyne + RNH$_2$ ⟶ N-substituted anilines

(*iv*) **Dimerisation.** In the absence of any nucleophile, benzyne undergoes dimerisation

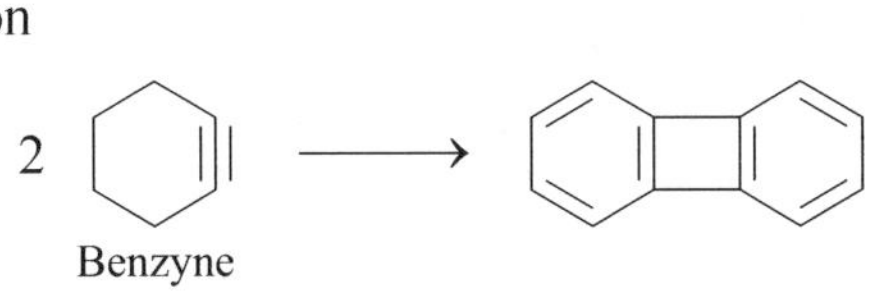

2 Benzyne ⟶

(*v*) **Reaction with olefins**

Benzyne + CH$_2$=CHCN ⟶
Acrylo nitrile

See also section 2.8.2.2.5.

2.5 TYPES OF REAGENTS USED IN A CHEMICAL REACTION

We know that an organic reaction proceeds by the attack of a reagent on the substrate molecule. Under suitable conditions and the influence of an attacking reagent, the substrate molecule undergoes heterolysis to form reactive intermediates such as carbocations and carbanions, which combine with the reagent to give the final product. It is, therefore, desirable to understand the type of various attacking reagents.

The attacking reagents are of two types *viz.* (*i*) Electrophilic reagents and (*ii*) Nucleophilic reagents. These are briefly discussed below:

2.5.1 Electrophilic Reagents

These are electron loving species (*electro* = electron, *philic* = loving). These again are of two types:

(*a*) Positive Electrophiles

(*b*) Neutral Electrophiles

Positive Electrophiles (E$^+$). These electrophiles carry a positive charge and are deficient of two electrons. Typical example are carbocation and bromonium ion.

$$
\begin{array}{c}
H \\
| \\
R-\overset{+}{C} \\
| \\
H
\end{array}
\qquad\qquad
\overset{+}{:}\!\ddot{B}r\!:
$$

Carbocation   Bromonium ion

Both the carbocation and bromonium ion have six electron each in their outer shell and so are deficient of two electrons.

As we know, the positively charged electrophiles are produced by the heterolytic fission of neutral molecules. For example, bromonium ion is produced by the heterolysis of bromine molecule.

$$
:\ddot{B}r : \ddot{B}r: \xrightarrow{\text{Heterolysis}} :\overset{-}{\ddot{B}r} + \overset{+}{:\!\ddot{B}r\!:}
$$

Bromine molecule   Bromide ion   Bromonium ion

In a similar way, carbocations are formed by the heterolytic fission of carbon-halogen bonds as described earlier.

Neutral Electrophiles (E). In these electrophiles, the central atom has six electrons but carries no charge. Some typical examples are:

$$
\begin{array}{c}
F \\
| \\
F-B \\
| \\
F
\end{array}
\qquad\qquad
\begin{array}{c}
Cl \\
| \\
Cl-Al \\
| \\
Cl
\end{array}
$$

Boron trifluoride   Aluminium trichloride

Both types of electrophiles, being deficient of two electrons, react with substrates or the molecules which are electron rich and can donate an electron pair to the reagent.

Carbanion   Positive   (Product)
(substrate) electrophile

Carbanion   Neutral   (Product)
(substrate) electrophile

As is seen above, the positive electrophile gives neutral product while the neutral electrophile gives a product with a negative charge. Table 1.4 lists some common examples of positive and neutral electrophiles.

Table. Some common positive and neutral electrophiles

Positive electrophiles		Neutral electrophiles	
Carbocation	$-\overset{\displaystyle |}{\underset{\displaystyle |}{C}}$	Aluminium trichloride	$AlCl_3$
Bromonium ion	Br^+	Boron trifluoride	BF_3
Proton	H^+	Sulphur trioxide	SO_3
Hydronium ion	H_3O^+	Ferric chloride	$FeCl_3$
Nitronium ion	NO_2^+	Carbonyl group	$\diagdown C=O$
Sulphonium ion	$\overset{+}{S}O_2-OH$		
Diazonium ion	$C_6H_5N_2^+$		

2.5.2 Nucleophilic Reagents

These reagent are nucleus loving (*nucleo* = nucleus; *philic* = loving). These nucleophilic reagents are electron rich and are of two types:

(*a*) negative nucleophiles and (*b*) neutral neutrophiles.

Negative Nucleophiles (N̈ū). These nucleophiles have an excess electron pair and carry a negative charge. For example, carbanions and chloride ion:

$$R-\overset{\displaystyle H}{\underset{\displaystyle H}{\overset{|}{\underset{|}{C}}}}\!:^{\!-} \qquad\qquad :\ddot{\overset{..}{C}l:^{-}}$$

Carbanion   Chloride ion

Both carbanion and chloride ion have a spare electron pair each and carry a negative charge. Other example include hydroxide ion, $:\ddot{O}H^-$, cyanide ion, $^-C\equiv N$; bisulphide ion, $:\ddot{S}H^-$; alkoxy ion, $:\ddot{O}R^-$; hydride ion, $^-\ddot{H}$ and halide ion, $:\ddot{\overset{..}{X}}:^{-}$.

Neutral Nucleophiles (Nu:). These nucleophiles do have a pair of electrons but carry no charge. The common examples are ammonia and water.

$$H-\overset{\displaystyle H}{\underset{\displaystyle H}{\overset{|}{\underset{|}{N}}}}\!: \qquad\qquad H-\ddot{O}-H$$

Ammonia   Water

Other examples of neutral nucleophiles are: alcohols, $R-\ddot{O}-H$; thioalcohols, $R-\ddot{S}-H$ and ethers $R-\ddot{O}-R$.

As both the charged and neutral nucleophiles have a spare pair of electrons, they attack the substrate molecule which is defficient of two electrons. Thus,

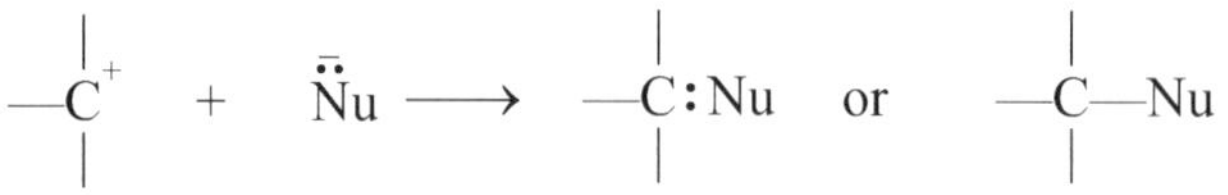

| Carbanion | Negative | | (Product) |
| (substrate) | nucleophile | | |

| Carbocation | Neutral | | (Product) |
| (substrate) | nucleophile | | |

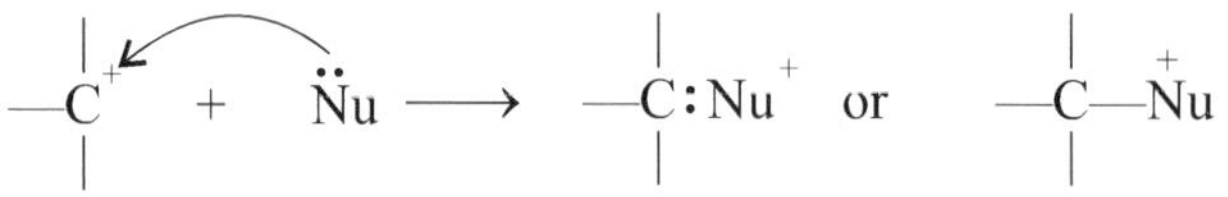

As is seen above, the charged nucleophile gives a neutral product while in case of neutral nucleophile, the nucleophile part develops a positive charge.

2.6   TYPE OF REACTIONS INVOLVED IN A CHEMICAL REACTION

Chemical reactions are known to take place in one step or a number of steps. The former type of reactions are called **concerted reactions** and the reactions of second type are called **step wise reactions.** In step wise reactions, a starting material is first converted to an unstable reactive intermediate, which forms the final product. Chemical reactions are of the following types:

- Addition reactions
- Substitution reactions
- Photochemical reactions
- Reduction reactions

- Elimination reactions
- Rearrangement reactions
- Oxidation reactions

2.6.1 Addition Reactions

In addition reactions two reactant molecules combine to form products, which contain atoms of both the reactants. Typical addition reactions are addition of water (hydration) and hydrogen (catalytic hydrogenation) to alkenes

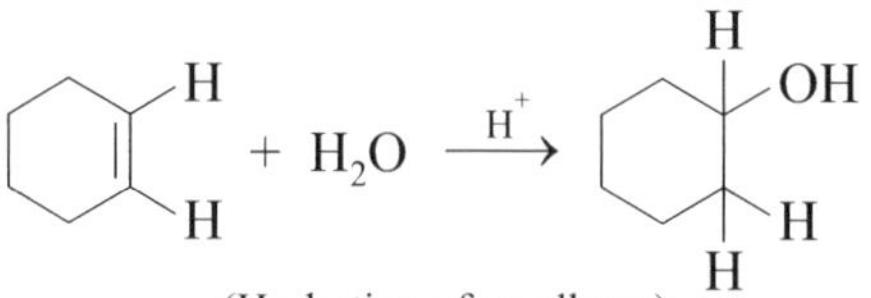

(Hydration of an alkene)

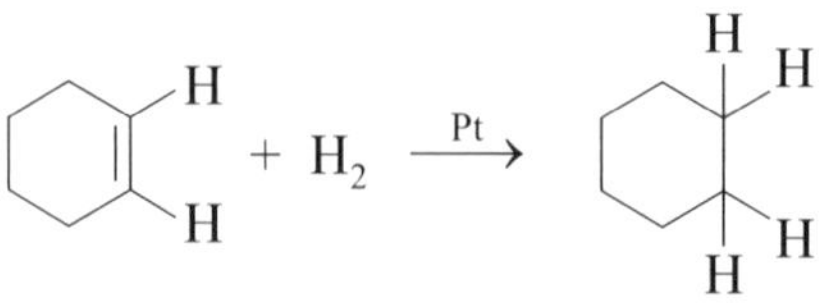

(Catalytic hydrogenation of an alkene)

As seen, in the hydration reaction, cyclohexene and water combine to form cyclohexanol and in the catalytic hydrogenation reaction, addition of hydrogen takes place to cyclohexene in presence of a metal catalyst to produce cyclohexane.

Some reactions require the presence of a substance called catalyst, which is not part of the product. However, catalyst facilitate the reaction.

The addition reactions are characteristic of compounds containing multiple bonds as in the case of alkenes and alkynes. However, as we will see subsequently, in compounds containing carbonyl group, addition reaction can also take place.

We know that a carbon-carbon double bond is made up of a stronger σ bond and a weak π bond, and that a carbon-carbon triple bond consists of one σ bond and two π bonds. Less energy is needed to break a π bond than a σ bond. It is found that an ethylenic π bond is about 15 kcal/more weaker than a σ bond. So it is expected that in an addition reaction to a carbon-carbon double bond (case of alkenes), the weaker π bond breaks open up leaving the σ bond intact. The addition of halogen to alkene is represented as follows:

$$\underset{\text{Alkene}}{\overset{\pi}{\underset{\sigma}{\text{C}=\text{C}}}} \;+\; X_2 \;\longrightarrow\; \underset{\text{Addition product}}{\overset{\text{X} \quad \text{X}}{\underset{\sigma}{\text{C}-\text{C}}}}$$

(Here, the straight line represents σ bond and the curved line represents π bond.)

Since a carbon-carbon triple bond contains two π bonds and one σ bond, the addition of a reagent will take place in two steps.

$$\underset{\text{Alkyne}}{\overset{\pi}{\underset{\pi}{\text{C}\underset{\sigma}{=}\text{C}}}} \;\overset{X_2}{\longrightarrow}\; \underset{\text{First adduct}}{\overset{\text{X} \quad \text{X}}{\underset{\pi}{\text{C}-\text{C}}}} \;\overset{X_2}{\longrightarrow}\; \underset{\substack{\text{X} \quad \text{X}\\ \text{Second adduct}}}{\overset{\text{X} \quad \text{X}}{\text{C}-\text{C}}}$$

Addition reactions to the carbon-carbon multiple bonds can be of the following types:

- Electrophilic additions
- Free radical additions
- Concerted additions
- Nucleophilic additions. These reactions are characteristic of carbonyl compounds

2.6.1.1 Electrophilic Addition Reactions

It is well known that unsaturated compounds like alkenes decolourise a solution of bromine in a suitable solvent like carbon tetrachloride. This is due to the formation of an addition product and the reaction is called addition reaction.

$$CH_2=CH_2 + Br—Br \longrightarrow CH_2—CH_2$$

Ethylene

$$\underset{Br\quad Br}{|\qquad |}$$

(Addition product)

It is believed that when an alkene comes in close proximity of a bromine molecule, it gets polarised due to π electrons of the alkene, thereby resulting in the formation of a dipole in the bromine molecule. The positive end of this polarised bromine molecule gets loosely attached to the π electron cloud of the alkene, resulting in the formation of a π complex. The π complex then breaks down to give a carbocation and bromide ion.

$$\begin{array}{cccc} Br—Br & \overset{\delta+}{\underset{}{Br}}—\overset{\delta-}{Br} & & \\ + & \uparrow & & \\ H_2C=CH_2 \longrightarrow & H_2C=CH_2 & \longrightarrow \overset{+}{H_2C}—CH_2Br + \overset{-}{Br} \\ & \pi\text{-Complex} & Carbocation \quad Bromide\ ion \end{array}$$

The carbocation thus produced has a transitory existence and combines readily with the nucleophilic bromine leading to the formation of addition product.

$$Br—CH_2—\overset{+}{CH_2} + \overset{..}{Br}{}^{-} \xrightarrow[\text{attack}]{\text{Nucleophilic}} Br—CH_2—CH_2—Br$$

Carbocation   Addition product

The above mechanism is supported by the observation that mixed products are obtained when an alkene is treated with bromine water in the presence of other nucleophiles such as chloride or nitrate ions. In these cases, step wise addition takes place and not the direct addition of a halogen molecule to an alkene. Thus,

$$Br—CH_2—\overset{+}{C}H_2 + Br^{-} \longrightarrow Br—CH_2—CH_2—Br$$

Carbocation   (Already present
in the solution)

$$Br—CH_2—\overset{+}{C}H_2 + Cl^{-} \longrightarrow Br—CH_2—CH_2—Cl$$

Carbocation   (Added to the
reaction mixture)

$$Br—CH_2—\overset{+}{C}H_2 + NO_3^{-} \longrightarrow Br—CH_2—CH_2—NO_3$$

Carbocation   (Added to the
reaction mixture)

As the above reaction is initiated by the positive end of the bromine dipole (electrophile), it is called **electrophilic addition reaction.**

It is known that the addition reactions of alkenes usually give *trans* addition products. The formation of *trans* products cannot be explained by the above reaction mechanism. It is now believed that the reaction proceeds *via* a cyclic intermediate (bromonium ion).

$$CH_2-\overset{\cdot\cdot}{Br}:$$
$$\overset{+}{CH_2} \longleftarrow$$

$\longrightarrow$

$$CH_2$$
$$\Big\rangle \overset{+}{Br}:$$
$$CH_2$$

Carbocation   Bromonium ion
                            (a cyclic intermediate)

Once the bromonium ion is formed, the attack of bromide ion can occur from the back side, giving the *trans* addition products; the attack from the front side is hindered due to the presence of bromine atom.

trans addition product   *cis* addition product

The most common electrophilic addition reactions to carbon-carbon multiple bonds include addition of hydrogen halides, water, halogen, hypocholous acid and addition to conjugated dienes.

(*i*) **Electrophilic addition of halogen halides (Hydrohalogenation):** We have see that the reaction of hydrogen halide to alkene gives alkyl halide. The order of reactivity of HX towards alkene is: HI > HBr > HCl > HF. In case of symmetrical alkene (like ethylene) only one product is obtained. However is case of unsymmetrical alkene like propene, in principle two products can be formed *i.e.*, 1-bromopropane and 2-bromopropane

$$CH_3CH=CH_2 + HBr$$
Propene

$$\longrightarrow CH_3CH-CH_2 \quad (\overset{H}{|} \cdots \overset{|}{Br})$$
2-Bromopropane

$$\longrightarrow CH_3CH-CH_2-Br \quad (\overset{|}{H})$$
1-Bromopropane

In practice, however, only one product, *viz.* 2-bromopropane is obtained.

On the basis of extensive work carried out by a Russian chemist Markovinkov on the addition of unsymmetrical reagents (like HCl, HBr, HI, H_2SO_4, HOCl, etc.) to unsymmetrical alkenes, an empirical rule known as **Markovnikov Rule** was formulated to predict formation of the addition product.

According to Markovnikov Rule, the positive end of the reagent becomes attached to the carbon atom of the double bond bearing larger number of H atoms. This is depicted as follows:

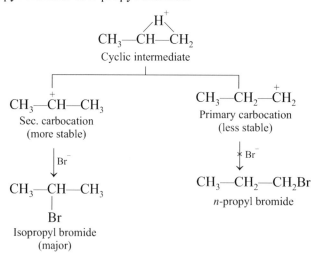

The formation of 2-bromopropane can be explained as follows:

Step 1. Propene on reaction with HBr gives a π complex, which in turn leads to the formation of a cyclic intermediate and Br⁻ ion.

Step 2. The cyclic intermediate can from two carbocations, *i.e.*, primary and secondary.

Step 3. The carbocations may then react with the nucleophile (Br⁻) to give either isopropyl bromide or *n*-propyl bromide.

Since we know that the stabilities of the carbocations are in the order $3° > 2° > 1°$, so in the above case the secondary carbocation being more stable than the primary carbocation, the product obtained is predominantly isopropyl bromide.

The formation of isopropyl bromide as the major product in the above reaction can also be explained by consideration of the electronic effect of the substituents on the double bond. Thus, in the case of propene, due to +I effect of the methyl group the π electrons are displaced towards the terminal carbon atom which acquires partial negative charge. So the proton adds on the carbon atom farthest form the methyl group. This is followed by the addition of the halide ion to the carbocation formed at the adjacent carbon atom.

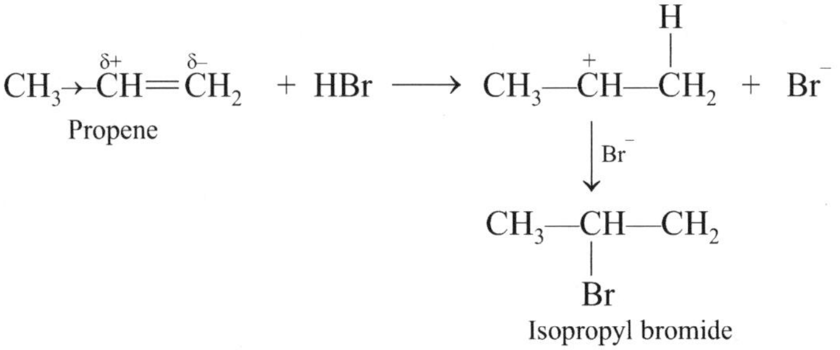

$$CH_3 \overset{\delta+}{\rightarrow} \overset{\delta-}{CH} = CH_2 \ + \ HBr \ \longrightarrow \ CH_3 - \overset{+}{CH} - \overset{\overset{\displaystyle H}{\displaystyle |}}{CH_2} \ + \ \overset{-}{Br}$$

Propene

$$\downarrow \overset{-}{Br}$$

$$CH_3 - \underset{\underset{\displaystyle Br}{\displaystyle |}}{CH} - CH_2$$

Isopropyl bromide

In a similar way, the reaction of HBr to 2-methylpropene gives 2-bromo-2-methylpropane as the major product; the alternative addition product, *viz.* 1-bromo-2-methyl-propane is not obtained.

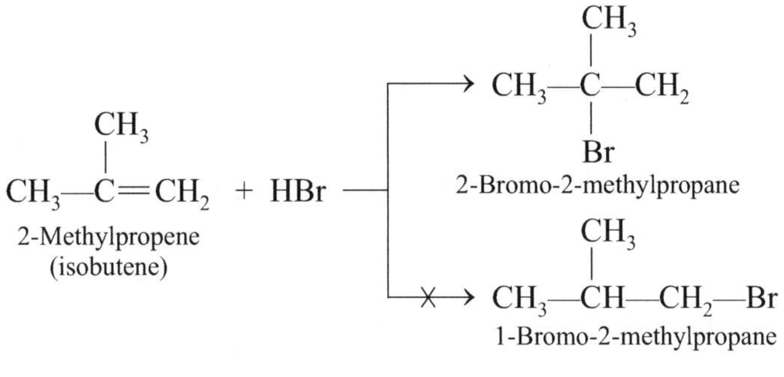

$$CH_3 - \underset{\underset{\displaystyle CH_3}{\displaystyle |}}{C} = CH_2 \ + \ HBr$$

2-Methylpropene
(isobutene)

$$\longrightarrow CH_3 - \underset{\underset{\displaystyle Br}{\displaystyle |}}{\overset{\overset{\displaystyle CH_3}{\displaystyle |}}{C}} - CH_2$$

2-Bromo-2-methylpropane

$$\overset{\times}{\longrightarrow} CH_3 - \overset{\overset{\displaystyle CH_3}{\displaystyle |}}{CH} - CH_2 - Br$$

1-Bromo-2-methylpropane

An interesting example in the addition of HBr to propenenitrile. In this case, the product obtained is 1-bromopropanenitrile (as the major product), which is obtained from the primary carbocation

$$N \equiv C \ CH = CH_2 \ \xrightarrow{HBr} \ NC \ \overset{+}{CH} - \overset{\overset{\displaystyle H}{\displaystyle |}}{CH_2} \ \xrightarrow{\overset{-}{Br}} \ NC \cdot \overset{\overset{\displaystyle Br}{\displaystyle |}}{CH} - \overset{\overset{\displaystyle H}{\displaystyle |}}{CH_2}$$

Propenenitrile

2° carbocation
(less stable)

2-Bromopropane
nitrile

$$\xrightarrow{HBr} \ NC \ \overset{\overset{\displaystyle H}{\displaystyle |}}{CH} - \overset{+}{CH_2} \ \xrightarrow{\overset{-}{Br}} \ NCCH - CH_2Br$$

1° carbocation
more stable

1-Bromopropane
nitrile

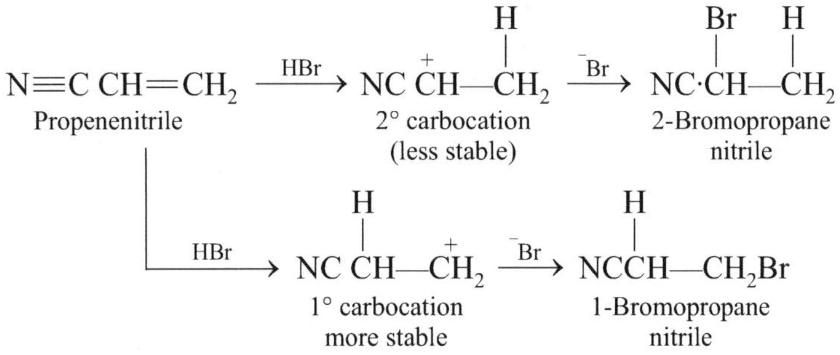

In the above case, the 1° carbocation is more stable. This is due to strong electron withdrawing effect of the nitrile group. Also the carbocation is separated from the CN group by two carbon atoms and destabilisation by the inductive effect is less.

In view of the examples cited above the Markownikoffs rule can be modified as follows.

"In the addition of hydrogen halide to alkenes, the more stable carbocation is formed which then adds to the negative ion to form the product". This rule predicts the orientation pattern of the addition of unsymmetrical reagents to unsymmetrical alkenes. It should, however be emphasised that this rule is applicable to only the addition of electrophilic reagents.

In case, in case of alkenes which have both the carbon atoms forming the double bond the same number of hydrogen atoms in an unsymmetrical alkene, a mixture of products is obtained. For example 2-pentene on reaction HBr gives a mixture of two products as shown below:

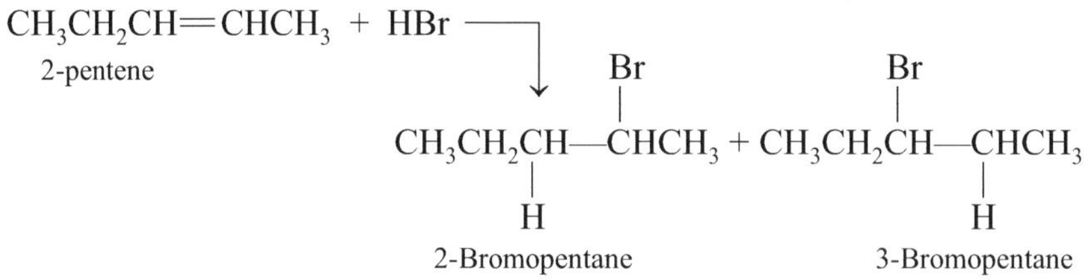

$$CH_3CH_2CH=CHCH_3 + HBr$$
2-pentene

$$\underset{\text{2-Bromopentane}}{CH_3CH_2CH\underset{|}{\overset{Br}{|}}CHCH_3} + \underset{\text{3-Bromopentane}}{CH_3CH_2\underset{|}{\overset{Br}{|}}CH\underset{H}{|}CHCH_3}$$

Alkynes also undergo hydrohalogenation as in case of alkenes. The addition is this case also is in accordance to Markownikoff's rule. Thus, ethyne on reaction with HBr first forms 1-bromoethene and then 1, 1-dibromoethane

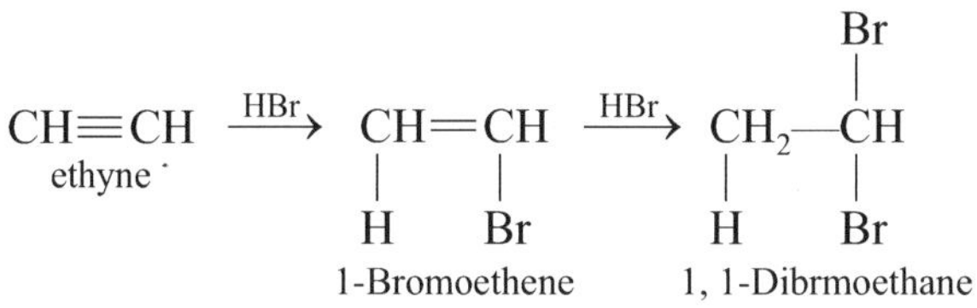

$$\underset{\text{ethyne}}{CH\equiv CH} \xrightarrow{HBr} \underset{\underset{H\quad Br}{|\quad|}}{CH=CH} \xrightarrow{HBr} \underset{\underset{H\quad Br}{|\quad|}}{CH_2-\overset{Br}{\overset{|}{CH}}}$$
1-Bromoethene   1, 1-Dibrmoethane

In this case also, the mechanism of the reaction is same as in case of hydrogenation of alkenes. Addition of one molecule of HBr gives bromoethene

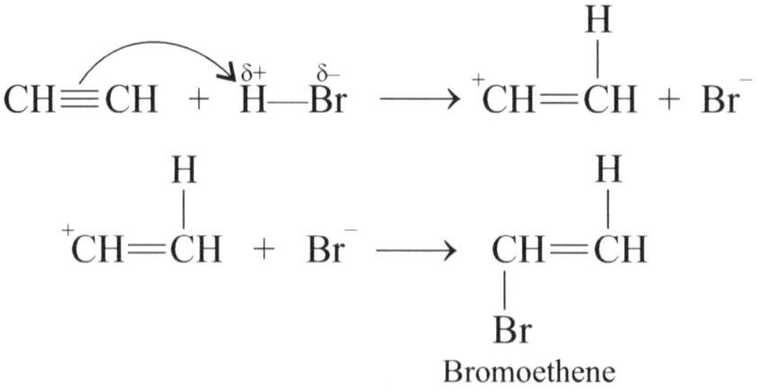

$$CH\equiv CH + \overset{\delta+}{H}-\overset{\delta-}{Br} \longrightarrow \overset{+}{CH}=\overset{\overset{H}{|}}{CH} + Br^-$$

$$\overset{+}{CH}=\overset{\overset{H}{|}}{CH} + Br^- \longrightarrow CH=\underset{\underset{Br}{|}}{CH}$$
Bromoethene

Subsequently, addition of another molecule of HBr could give either 2° carbocation or a 1° carbocation. As the 2° carbocation is more stable than the 1° carbocation, the reaction proceeds *via.* the 2° carbocation to give 1, 1-dibromoethane

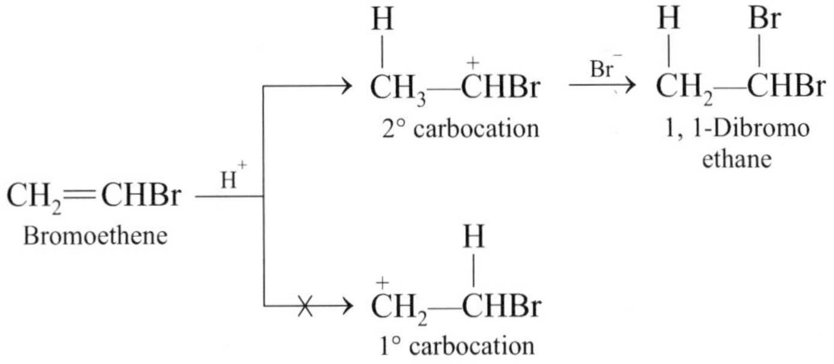

$$
\begin{array}{cc}
\underset{\substack{\text{2° carbocation}}}{\overset{\overset{H}{|}}{CH_3}-\overset{+}{CH}Br} & \xrightarrow{\overset{-}{Br}} \underset{\substack{\text{1, 1-Dibromo}\\\text{ethane}}}{\overset{\overset{H}{|}}{CH_2}-\overset{\overset{Br}{|}}{CH}Br}
\end{array}
$$

$$
\underset{\text{Bromoethene}}{CH_2=CHBr} \xrightarrow{\overset{+}{H}}
$$

$$
\times\rightarrow \underset{\substack{\text{1° carbocation}}}{\overset{\overset{H}{|}}{\overset{+}{CH_2}}-CHBr}
$$

(*ii*) **Addition of Water (Hydration).** Addition of water to carbon-carbon double bond takes place in presence of a mineral acid. This reaction is called hydration of alkene. Some examples are given below:

$$
\underset{\text{Ethylene}}{CH_2=CH_2} + \overset{\delta+}{H}-\overset{\delta-}{OH} \xrightarrow{H_2SO_4} \underset{\text{Ethyl alcohol}}{CH_3CH_2OH}
$$

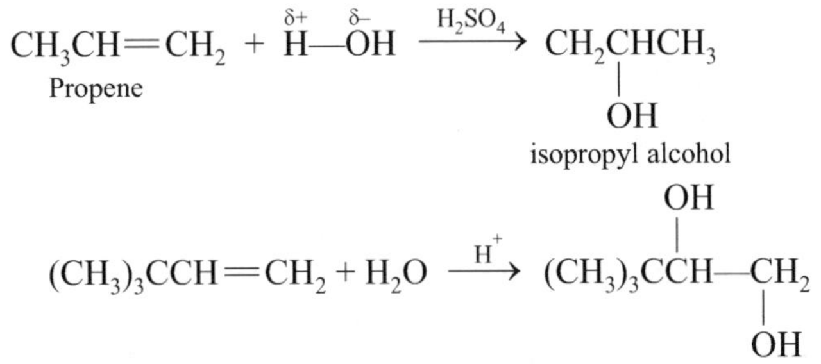

$$
\underset{\text{Propene}}{CH_3CH=CH_2} + \overset{\delta+}{H}-\overset{\delta-}{OH} \xrightarrow{H_2SO_4} \underset{\substack{\text{isopropyl alcohol}}}{\overset{}{CH_2CHCH_3}}
\\ \qquad\qquad\qquad\qquad\qquad\qquad\qquad\; |
\\ \qquad\qquad\qquad\qquad\qquad\qquad\quad OH
$$

$$
(CH_3)_3CCH=CH_2 + H_2O \xrightarrow{\overset{+}{H}} \underset{}{(CH_3)_3\overset{}{CCH}-CH_2}
\\ \qquad\qquad\qquad\qquad\qquad\qquad\qquad\qquad\qquad\quad |
\\ \qquad\qquad\qquad\qquad\qquad\qquad\qquad\qquad\quad OH
$$

The hydration of alkene also follows Markownikoff's rule. The reaction occurs in two steps as shown below:

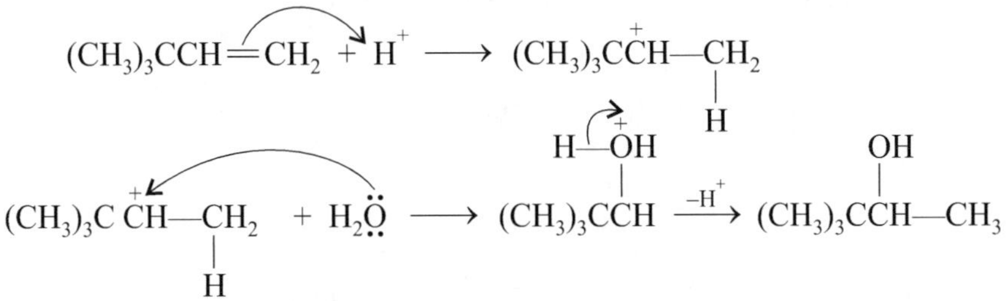

$$
(CH_3)_3CCH=CH_2 + \overset{+}{H} \longrightarrow (CH_3)_3\overset{+}{CCH}-CH_2
$$

$$
\underset{\overset{|}{H}}{(CH_3)_3\overset{+}{C}\ CH-CH_2} + H_2\overset{..}{\overset{..}{O}} \longrightarrow (CH_3)_3\overset{\overset{H-OH}{\downarrow}}{CCH} \xrightarrow{-\overset{+}{H}} \underset{\overset{|}{OH}}{(CH_3)_3CCH-CH_3}
$$

In this reaction a carbocation is involved and so rearrangement is possible. The secondary carbocation can undergo a 1, 2-shift of methyl group to yield a more stable tertiary carbocation, which reacts with water to form the alcohol.

$$CH_3—\overset{\overset{\displaystyle CH_3}{|}}{\underset{\underset{\displaystyle CH_3}{|}}{C}}—\overset{+}{C}HCH_3 \quad \xrightarrow{CH_3 \text{ shift}} \quad CH_3—\overset{\overset{\displaystyle CH_3}{|}}{\underset{\underset{\displaystyle CH_3}{|}}{\overset{+}{C}}}—CH—CH_3$$

2° carbocation   3° carbocation
(less stable)   (more stable)

$$\Big\downarrow \begin{matrix} H_2O \\ -H^+ \end{matrix}$$

$$CH_3—\overset{\overset{\displaystyle OH}{|}}{\underset{\underset{\displaystyle CH_3}{|}}{C}}—\overset{\overset{\displaystyle CH_3}{|}}{C}HCH_3$$

Markownikoff's hydration of an alkene can also be accomplished by
Oxymercuration-demercuration. The method involves reaction of the
alkene will mercuric acetate in presence of water to give hydroxy-mercural
compound, which on reduction with $NaBH_4$ produces an alcohol. This
method gives better yields than acid catalyst hydration.

$$CH_3CH_2CH{=}CH_2 + H_2O + Hg(O\overset{\overset{\displaystyle O}{\|}}{C}CH_3)_2 \longrightarrow CH_3CH_2\overset{\overset{\displaystyle OH}{|}}{\underset{\underset{\displaystyle HgO_2CH_3}{|}}{C}H}—CH_2$$

1-Butene   Hydroxy-mercural compound

$$\xrightarrow{NaBH_4} CH_3CH_2\overset{\overset{\displaystyle OH}{|}}{\underset{\underset{\displaystyle H}{|}}{C}H}—CH_2$$

2-Butanol

The mechanism of the reaction involves the formation of the electrophile,
$\overset{+}{H}gO_2CCH_3$ which attacks the carbon-carbon double bond. Various steps
involved are given below:

$$Hg(O_2CCH_3)_2 \rightleftharpoons Hg^+O_2CCH_3 + \bar{O}_2CCH_3$$

Mercuric acetate

$$\overset{}{\underset{}{>}}C{=}C\overset{}{\underset{}{<}} \;+\; H\overset{+}{g}O_2CCH_3 \longrightarrow \overset{\delta+}{\underset{\underset{\underset{\delta-}{HgO_2CCH_3}}{|}}{C}}—C$$

$$\overset{\delta+}{\underset{\underset{\underset{\delta-}{HgO_2CCH_3}}{|}}{C}}—C \xrightarrow{H_2\ddot{O}} —\overset{\overset{\displaystyle \overset{+}{O}H_2}{|}}{\underset{\underset{\displaystyle HgO_2CCH_3}{|}}{C}}—C— \xrightarrow{-H^+} —\overset{\overset{\displaystyle OH}{|}}{\underset{\underset{\displaystyle HgO_2CCH_3}{|}}{C}}—C—$$

Oxymercuration is regiospecific and stereospecific (anti hydroxylation). This is attributed to the formation of cyclic intermediate.

The alkynes on hydration with water in presence of $H_2SO_4/HgSO_4$ give carbonyl compounds (acetaldehyde from ethyne) *via.* the formation of enol

$$CH\equiv CH + H_2O \xrightarrow{H_2SO_4,\ HgSO_4} CH_2\!\!=\!\!CH_2 \rightleftharpoons CH_3\!-\!CH$$

Ethyne   OH   O

                                  enol   Acetaldehyde

$$CH_3\!-\!CH\equiv CH + H_2O \xrightarrow{H_2SO_4,\ HgSO_4} CH_3\!-\!C\!\!=\!\!CH_2 \rightleftharpoons CH_3\!-\!C\!-\!CH_3$$

Propyne   OH   O

                                          Propanone

                                        (ketone)

(*iii*) **Addition of Halogen (Halogenation).** Alkenes on treatment with halogens given 1, 2-dihalogenated alkenes or dihalides. The effective electrophiles are bromine and chlorine. This is the best method for the preparation of vicinal dihalides.

$$CH_2\!\!=\!\!CH_2 + Br_2 \xrightarrow{CCl_4} CH_2\!-\!CH_2$$

ethene   Br   Br

                                1, 2-Dibromoethane

           CH$_3$ CH$_3$

$$CH_3\!-\!C\!\!=\!\!CH_2 + Cl_2 \xrightarrow{CCl_4} CH_3\!-\!C\!-\!CH_2$$

2-Methyl propene   Cl   Cl

                               1, 2-Dichloro-2-methyl

                                    propane

Fluorine and iodine do not react with carbon-carbon double or carbon-carbon triple bond. In fact, fluorine undergoes explosive reaction with alkene or alkyne. So special techniques are required for reaction with fluorine and iodine.

The reaction with alkenes are carried out in inert solvent like CCl_4. At higher temperature substitution products may be formed and so the reactions are carried out at room temperature.

Alkynes, like alkenes also react with chlorine and bromine to yield tetrahaloalkanes. Two molecules of halogens are required for reaction with a triple bond. A dihaloalkene is formed as an intermediate, which can also be isolated using appropriate conditions. Thus, ethyne reacts with bromine water to give only 1, 2-dibromoethene whereas with bromine alone, it forms 1, 1, 2, 2-tetrabromo ethane

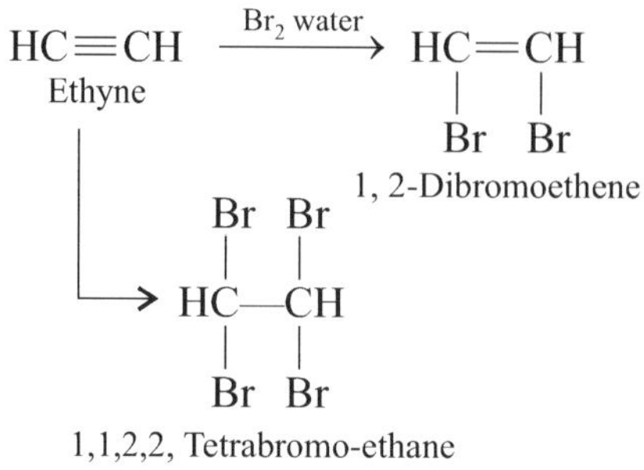

$$HC\equiv CH \xrightarrow{\text{Br}_2 \text{ water}} HC=CH$$

Ethyne

$$\underset{\text{Br}\quad\text{Br}}{HC=CH}$$

1, 2-Dibromoethene

$$\longrightarrow \underset{\overset{|}{\text{Br}}\quad\overset{|}{\text{Br}}}{\overset{\overset{\text{Br}}{|}\quad\overset{\text{Br}}{|}}{HC-CH}}$$

1,1,2,2, Tetrabromo-ethane

The addition of halogens to alkyne is stereoselective; the predominant product is the trans isomer

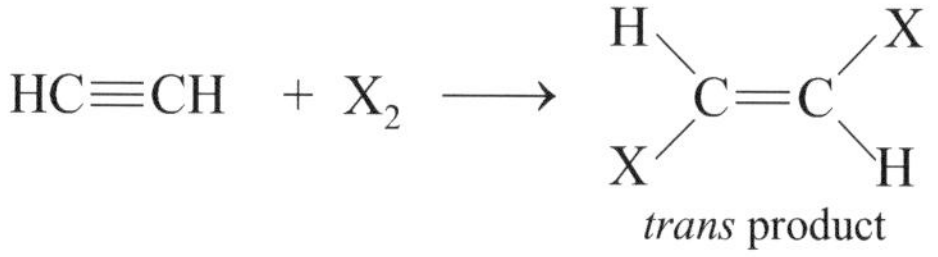

$$HC\equiv CH + X_2 \longrightarrow \underset{X}{\overset{H}{\diagdown}}C=C\underset{H}{\overset{X}{\diagup}}$$

trans product

The reaction of an alkene with bromine in carbon tetrachloride decolourise the yellow colour of bromine. This conforms the presence of carbon-carbon double bond.

The halogenation of an alkene gives *trans* addition product. The formation of *trans* addition product can be explained as follows. It is believed that the reaction takes place *via.* a cyclic intermediate.

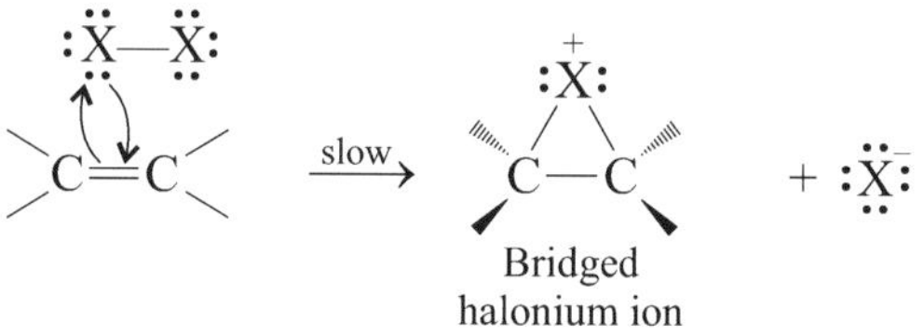

Bridged
halonium ion

The three membered ring containing a positively charged halogen is called a bridged halonium ion. In this three membered ring being strained and unstable, the ring opening takes place by the nucleophilic attack of X⁻ to give the addition product.

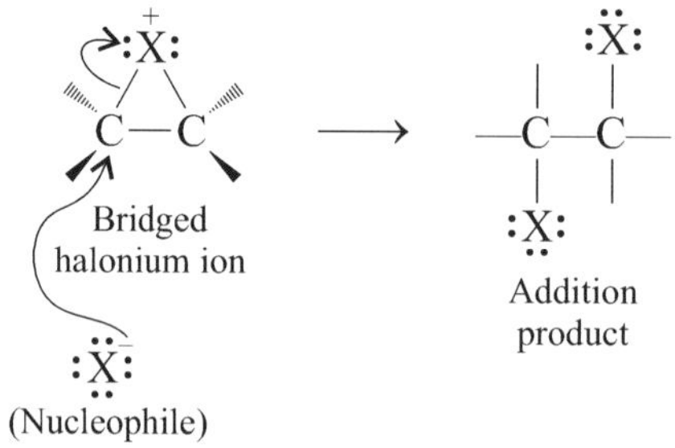

Bridged
halonium ion

:Ẍ:
(Nucleophile)

Addition
product

As seen, the attack of the nucleophile (:Ẍ:⁻) takes place from the back side giving *trans* addition product. The attack from the front side is hindered due to the presence of halogen atom.

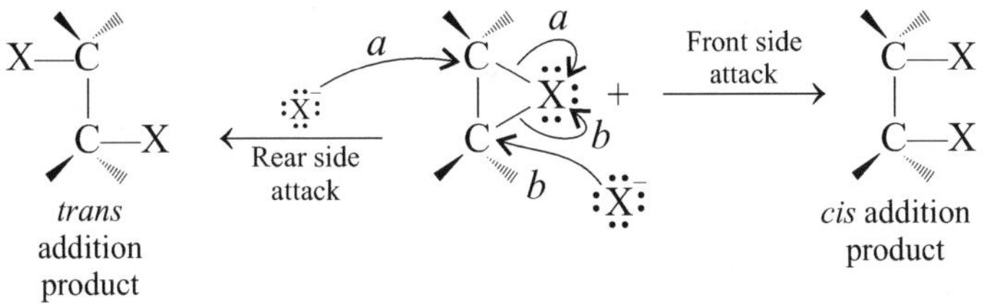

The halogenation of alkene is possible only with Br_2 and Cl_2. This is because I_2 reacts very slowly and F react with explosive violence.

Addition of HO–Cl (Formation of Halohydrins)

A solution of halogen (Br_2 or Cl_2) in water adds on to the alkenes to yield halohydrins (vicinial)

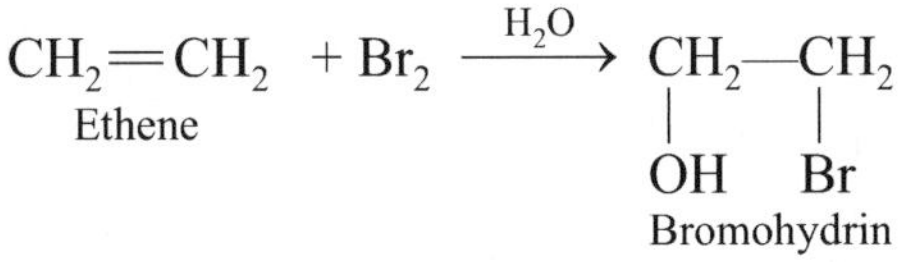

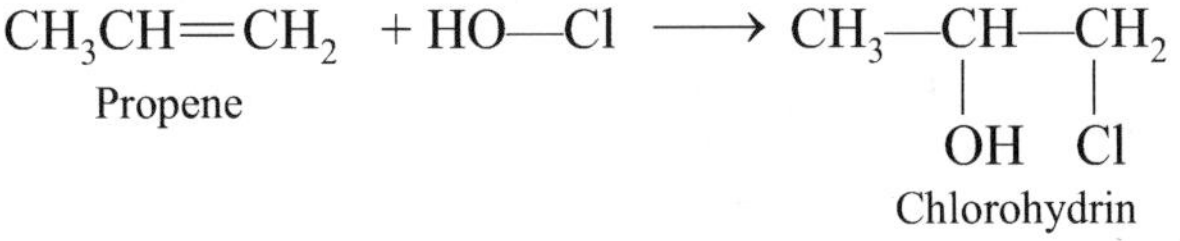

$$CH_2{=}CH_2 + Br_2 \xrightarrow{\ H_2O\ } \underset{\underset{\text{Bromohydrin}}{OH\quad Br}}{CH_2{-}CH_2}$$

Ethene

$$CH_3CH{=}CH_2 + HO{-}Cl \longrightarrow \underset{\underset{\text{Chlorohydrin}}{OH\quad Cl}}{CH_3{-}CH{-}CH_2}$$

Propene

In case of higher alkenes, like propene, the addition takes place as per Markownikoff's rule.

The addition of electrophile ($:\ddot{X}:^-$) to the π bond of the alkene gives a bridge halonium ion (as in the case of reaction of ethene with halogen) followed by reaction with water to give adduct as shown below:

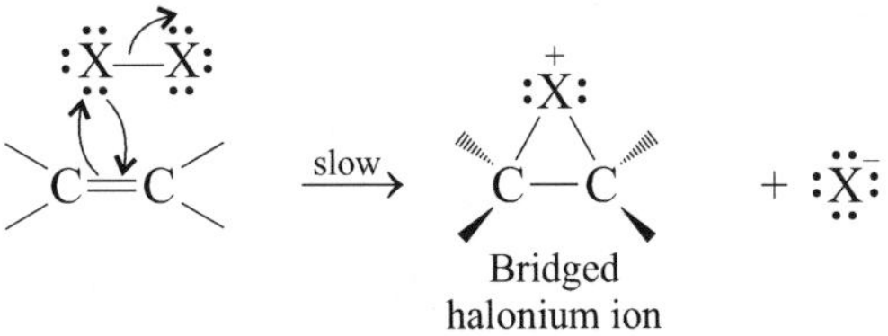

Bridged
halonium ion

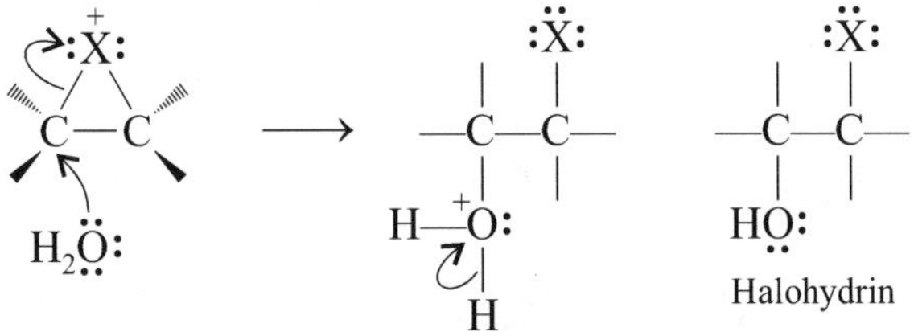

Halohydrin

The addition of X and OH across the carbon-carbon double bond is anti; since the two new bonds are formed in separate steps.

Bromohydrins are similarly formed from alkene by reacting with Br_2 and H_2O. These can also be formed by the reaction of alkenes with N-bromosuccinimide in aqueous DMSO. In water, NBS decomposes to form Br_2, which then reacts with alkene to form bromohydrin by the same mechanism

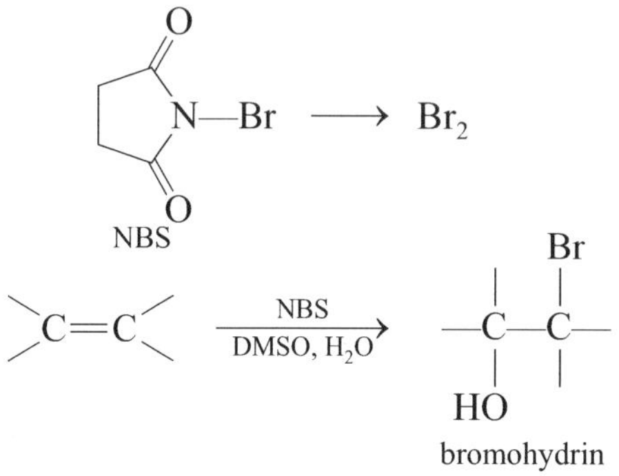

$$\text{NBS} \longrightarrow Br_2$$

bromohydrin

Addition to Conjugated Dienes

In isolated dienes, both the double bonds react independently. As an example 1, 4-pentadiene reacts with one mole of bromine to give 4, 5-dibromo-1-pentene

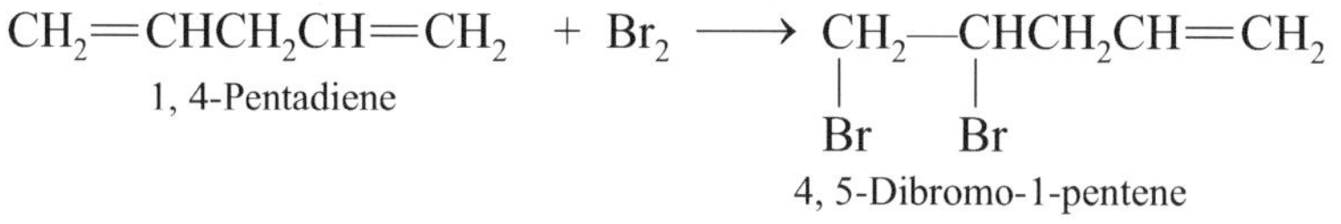

$$CH_2=CHCH_2CH=CH_2 + Br_2 \longrightarrow CH_2-CHCH_2CH=CH_2$$

1, 4-Pentadiene

Br   Br

4, 5-Dibromo-1-pentene

However, conjugated dienes behave differently compared to isolated dienes. In fact, conjugated diene, *e.g.*, 1, 3-butadiene undergoes electrophilic addition reactions under usual conditions forming a mixture of 1, 2- and 1, 4-addition products. The 1, 4-addition product predominates

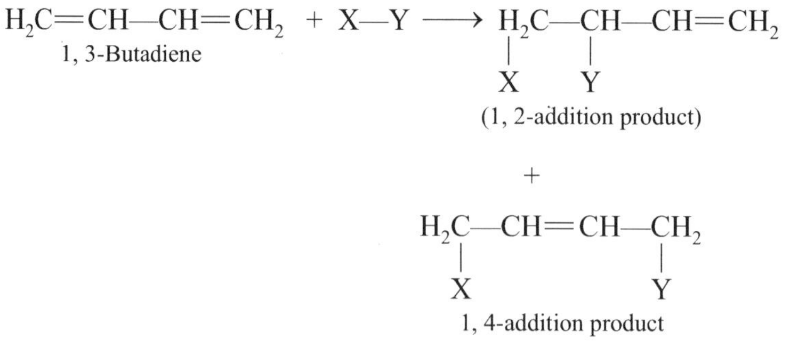

$$H_2C=CH-CH=CH_2 + X-Y \longrightarrow H_2C-CH-CH=CH_2$$

1, 3-Butadiene

X   Y

(1, 2-addition product)

+

$$H_2C-CH=CH-CH_2$$

X   Y

1, 4-addition product

The addition of bromine on butadiene yields a mixture of 3, 4-dibromo-1-butene (1, 2-addition product) and 1, 4-dibromo-2-butene (1, 4-addition product)

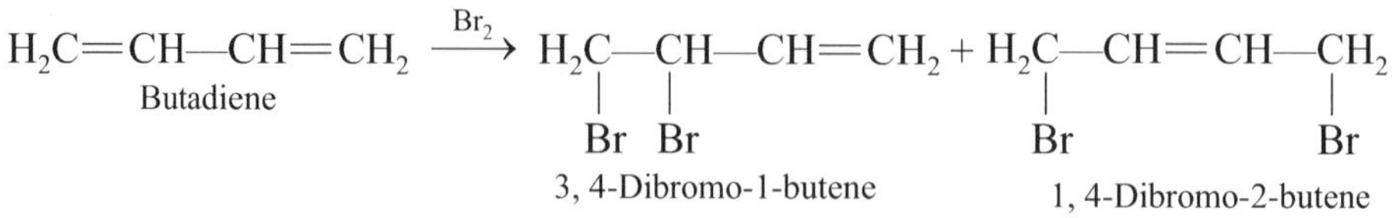

$$H_2C=CH-CH=CH_2 \xrightarrow{Br_2} H_2C-CH-CH=CH_2 + H_2C-CH=CH-CH_2$$

Butadiene

Br Br   Br   Br

3, 4-Dibromo-1-butene   1, 4-Dibromo-2-butene

In case of butadiene (a conjugated diene) the formation of 1, 4-addition product is characteristic of butadiene. This can be explained by assuming that initially the electrophile can attack either to C_1 or to C_2. As the addition of the electrophile at C_2 will give an unstable primary carbocation, the electrophile attack takes place at C_1 resulting in the formation of a resonance stabilized allylic cation, which adds on to bromine to give a mixture of 3, 4-dibromo-1-butene and 1, 4-dibromo-2-butene.

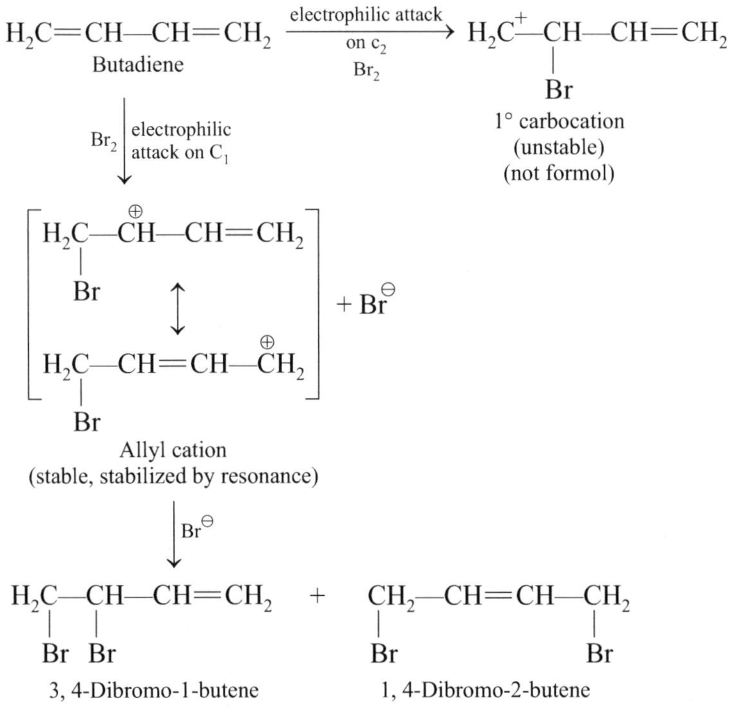

$$H_2C=CH—CH=CH_2 \xrightarrow[\substack{\text{on } c_2 \\ Br_2}]{\text{electrophilic attack}} H_2\overset{+}{C}—CH—CH=CH_2$$

Butadiene

$$\underset{Br}{\overset{|}{}}$$
1° carbocation
(unstable)
(not formol)

Br_2 electrophilic attack on C_1

$$\left[\begin{array}{c} H_2C—\overset{\oplus}{CH}—CH=CH_2 \\ \underset{Br}{\overset{|}{}} \quad \updownarrow \\ H_2C—CH=CH—\overset{\oplus}{CH_2} \\ \underset{Br}{\overset{|}{}} \end{array}\right] + \overset{\ominus}{Br}$$

Allyl cation
(stable, stabilized by resonance)

$$\Big\downarrow Br^{\ominus}$$

$$\underset{Br\ Br}{H_2C—CH—CH=CH_2} \quad + \quad \underset{Br}{CH_2—CH=CH—CH_2} \quad \underset{Br}{}$$

3, 4-Dibromo-1-butene   1, 4-Dibromo-2-butene

In the intermediate allyl cation, the positive charge is on either of the two carbons (as shown above). Both these positions may be attacked by the nucleophile to give 1, 2- and the 1, 4-addition product.

Similarly, addition of hydrogen chloride to butadiene gives a mixture of α-methyl allyl chloride and crotyl chloride.

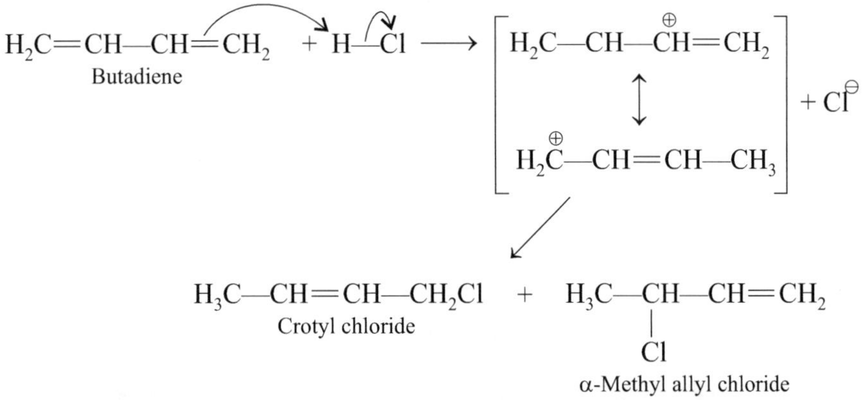

$$H_2C=CH—CH=CH_2 \quad + H—Cl \longrightarrow \left[\begin{array}{c} H_2C—CH—\overset{\oplus}{CH}=CH_2 \\ \updownarrow \\ H_2\overset{\oplus}{C}—CH=CH—CH_3 \end{array}\right] + Cl^{\ominus}$$

Butadiene

$$H_3C—CH=CH—CH_2Cl \quad + \quad \underset{Cl}{H_3C—CH—CH=CH_2}$$

Crotyl chloride   α-Methyl allyl chloride

The ratio of the two product formed depends on the temperature. It is found that at lower temperature ($-80°C$), the 1, 2-product premoninates while at higher temperature ($40°C$) the 1, 4-addition product is the major product. Though both the isomeric products are stable at low temperature, prolonged heating of either of them yield the same equilibrium mixture in which 1, 4-product predominate. Thus, 1, 2-product is kinetically controlled and the 1, 4-adduct is thermodynamically controlled.

2.6.1.2 Free Radical Addition Reactions

We have seen that the reaction of propene with alkyl halide (*e.g.*, HBr) gives isopropyl bromide as per Markownikoff's rule. It was found by Kharash and Mayo that in the presence of peroxide, the product formed was contrary to Markownikoff's rule.

This phenomenon is called **Kharash or Peroxide Effect.** Thus, the reaction of propene with HBr in the presence of peroxide gives *n*-propyl bromide. However in the absence of peroxide, the major product is isopropyl bromide.

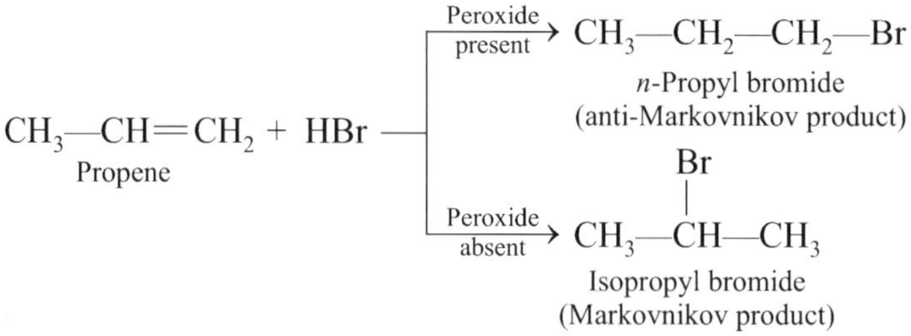

$$CH_3—CH=CH_2 + HBr$$

Peroxide present → $CH_3—CH_2—CH_2—Br$
n-Propyl bromide
(anti-Markovnikov product)

Peroxide absent →
$$CH_3—\overset{\overset{\displaystyle Br}{|}}{CH}—CH_3$$
Isopropyl bromide
(Markovnikov product)

The formation of anti-Markovnikov addition product is explained by the free radical mechanism rather than the ionic mechanism given for Markovnikov addition product. The step involved in the free radical mechanism are discussed below:

(*a*) Firstly, the peroxide dissociates into two free radicals called alkoxy radicals. These, in turn, attack HBr to form bromine free radical.

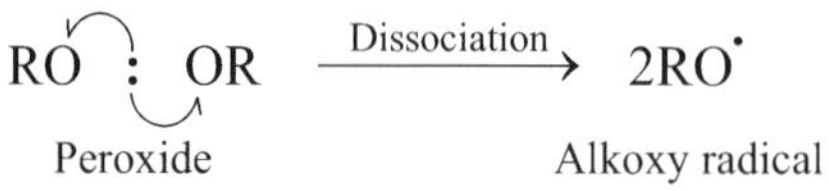

$$RO : OR \xrightarrow{\text{Dissociation}} 2RO^•$$
Peroxide   Alkoxy radical

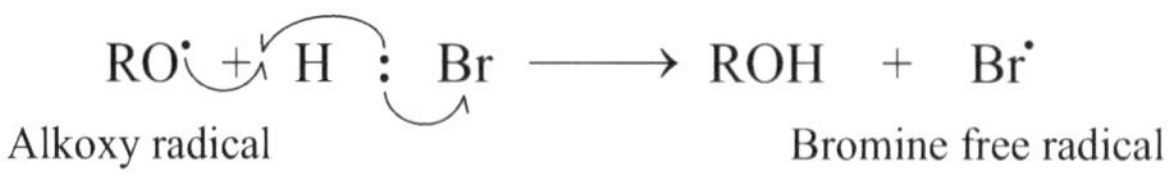

$$RO + H : Br \longrightarrow ROH + Br^•$$
Alkoxy radical   Bromine free radical

(*b*) In the second step, the bromine free radical attacks the alkene molecule to give two possible alkyl free radicals. Thus,

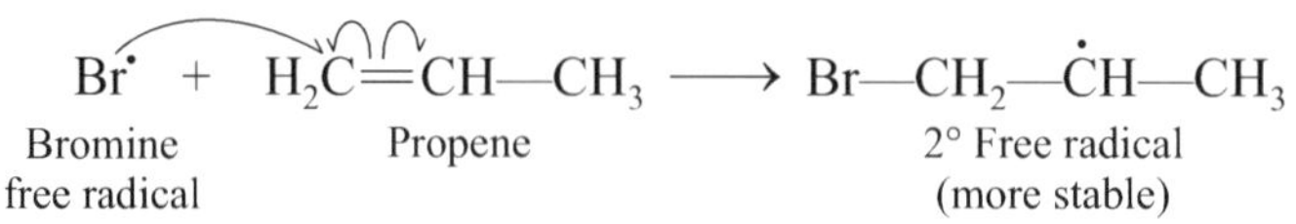

$$Br^• + H_2C=CH—CH_3 \longrightarrow Br—CH_2—\overset{•}{C}H—CH_3$$
Bromine
free radical   Propene   2° Free radical
(more stable)

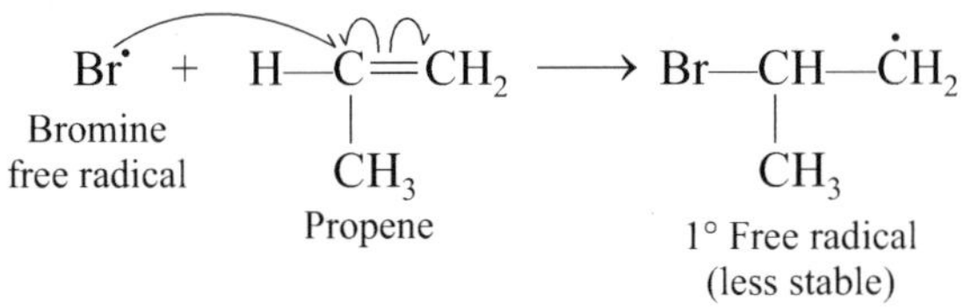

$$\overset{\frown}{Br} + H\overset{\curvearrowleft\curvearrowleft}{-\overset{|}{C}=CH_2} \longrightarrow Br-CH-\overset{\bullet}{C}H_2$$

Bromine		CH₃	CH₃
free radical		Propene	1° Free radical
			(less stable)

Br˙ + H—Ċ=CH₂ (CH₃) → Br—CH—ĊH₂ (CH₃)

Bromine free radical — Propene — 1° Free radical (less stable)

The order of stability of free radicals being $3° > 2° > 1°$, the $2°$ free radical is predominantly formed in this step.

(*c*) The more stable $2°$ free radical then reacts with HBr, forming anti-Markovnikov addition product and another bromine free radical which, in turn, propagates the chain.

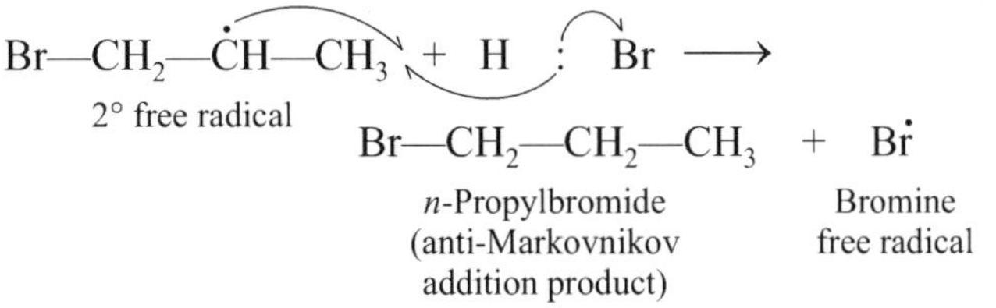

$$Br-CH_2-\overset{\bullet}{C}H-CH_3 + H \overset{\frown}{\,:\,} Br \longrightarrow$$

2° free radical

$$Br-CH_2-CH_2-CH_3 + Br^{\bullet}$$

n-Propylbromide	Bromine
(anti-Markovnikov	free radical
addition product)	

It should be noted that only HBr gives the anti-Markovnikov product. HCl and HI do not give anti-Markovnikov products in the presence of peroxides. This is because HCl and HI do not undergo free radical addition. In case of HCl, the cleavage of H—Cl bond to give Cl in step (*a*) is unfavourable as H—Cl bond is stronger than H—Br bond. In the case of HI, the free iodine radical (I.) is easily obtained as H—I bond is quite weak. However, the iodine radical immediately recombine to form I_2 molecule and, so, is not available for the reaction.

We have earlier seen (see page 87) that addition of HBr to propene nitrite gives 1-bromopropane nitrile as the major product. The same product is also obtained in the presence of peroxide.

$$CH_2=CH-CN \xrightarrow{HBr} \overset{+}{C}H_2-CH_2-CN \longrightarrow BrCH_2-CH_2-CN$$

Propene nitrite	1° carbocation	1-Bromopropane nitrile

$$\xrightarrow[HBr]{Peroxide} CH_2-\overset{\bullet}{C}H-CN \longrightarrow CH_2-CH_2-CN + Br^{\bullet}$$
$$\qquad\qquad\quad |\qquad\qquad\qquad\quad |$$
$$\qquad\qquad\quad Br\qquad\qquad\qquad Br$$

1-Bromopropane nitrile

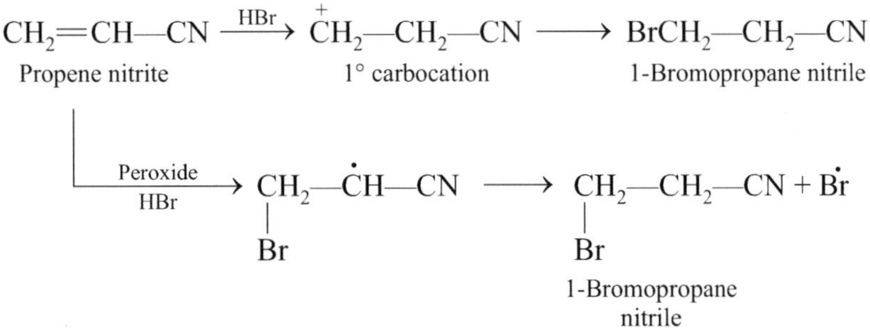

The halogenation of alkenes was shown earlier by treatment with halogen (Br_2 or Cl_2) in CCl_4 solution. Halogenation of alkenes can also be carried out in presence of light or peroxides, which follows radical mechanism as shown below:

$$X_2 \xrightarrow{h\nu} 2\overset{\bullet}{X}$$

An commercial process involving addition of free radical across double bonds is the reaction of benzene with chlorine in presence of light. Of the theoretically possible eight isomer, only three *viz.*, α, β and γ are formed in reasonable amounts. The γ-isomer, called gammexane is a potent insecticide

In this reaction chlorine radical is formed by homolytic fission of Cl_2.

2.6.1.3 Concerted Addition Reactions

A concerted reaction is a one step reaction. No matter how many bond are broken or formed. A starting material in a concerted reaction is directly converted into the product. Some important examples of concerted additions are hydroboration Diels-Alder reaction, ozonolysis and hydroxylation.

2.6.1.3.1 Hydroboration

Addition of borane (BH_3) to alkanes gives alkyl boranes while addition to an alkyne gives alkenyl borane.

The alkyl or alkenyl boranes are called organo boranes. In this process a new carbon hydrogen bond and a new carbon-boron bond are formed.

Organoboranes were discovered by Herbert C. Brown, who was awarded a Nobel prize in 1979 for his work with organoboron compounds.

Borane (BH_3) itself is unknown but its dimer, diborane (B_2H_6) behave as it were the hypothetical monomer.

The addition of borone to alkene takes place in a step wise fashion *via.* successive addition of each boron hydrogen bond to the alkene. The sequence of reaction is called **Hydroboration.**

Step (i) $CH_2{=}CH_2$ + $\underset{\overset{|}{H}}{\overset{\overset{H}{|}}{B}}{-}H$ $\longrightarrow$ $\underset{\text{Alkyl borane}}{H{-}B{-}CH_2CH_3}$ with H above B

Step (ii) $CH_2{=}CH_2$ + $H{-}\underset{}{\overset{\overset{H}{|}}{B}}{-}CH_2CH_3$ $\longrightarrow$ $\underset{\text{Dialkyl borane}}{H{-}\overset{\overset{CH_2CH_3}{|}}{B}{-}CH_2CH_3}$

Step (iii) $CH_2{=}CH_2$ + $H{-}\overset{\overset{CH_2CH_3}{|}}{B}{-}CH_2CH_3$ $\longrightarrow$ $\underset{\underset{\text{Trialkyl borane}}{\overset{|}{CH_2CH_3}}}{\overset{\overset{CH_2CH_3}{|}}{B}{-}CH_2CH_3}$

The addition of BH_3 ($H{-}BH_2$) to an alkene involves the addition of hydrogen (which is the electronegative proton of the molecule) to the carbon of the alkene which is more substituted. The net result is anti-Markownikoff's addition

$$\underset{\underset{\delta-\quad\delta+}{H{-}BH_2}}{CH_2CH{=}CH_2} \longrightarrow \underset{\underset{H\quad BH_2}{|\quad\ |}}{CH_2CH{-}CH_2} \qquad \text{...(1)}$$

The ease of substitution decrease with the increase in the alkyl substitutes on the double bond. Thus, trisubstituted alkene forms dialkyl borane and tetrasubstituted alkene form only monoalkylboranes.

$$\underset{\underset{\text{alkene}}{\text{Trisubstituted}}}{(CH_3)_2C{=}CHCH_3} \xrightarrow{(BH_3)_2} \underset{\text{Dialkyl borane}}{(CH_3)_2CH{-}\overset{\overset{CH_3}{|}}{CH}{-}BH{-}\overset{\overset{CH_3}{|}}{CH}{-}CH(CH_3)_2}$$

$$\underset{\underset{\text{alkene}}{\text{Tetrasubstituted}}}{(CH_3)_2C{=}C(CH_3)_2} \xrightarrow{(BH_3)_2} \underset{\underset{\text{Monosubstituted borane}}{\overset{|}{CH_3}}}{(CH_3)_3CH{-}\overset{\overset{CH_3}{|}}{C}{-}BH_2}$$

As seen (equation (*i*)), in hydrobroation the boron and the hydride ion add to the two carbon atoms the alkene simultaneously. This means that both B and H add from the same side of the double bond. In other words, the addition is *cis*-additions or syn-addition.

Subsequent oxidation of the organoborane gives an alcohol ($BH_2 \longrightarrow OH$) in which the hydroxyl group ends up in the same position as the boron atom. This means that there is retention of configuration at the carbon atom carrying the BH_2 group

organoborane   alcohol

The organoboranes are versatile intermediate for organic synthesis. Some typical examples are given below:

(i) Oxidation of organoboranes with alkaline H_2O_2 gives the corresponding alcohol. The overall result is the anti-Markownikoff's addition of water to a double bond.

$$3CH_3CH_2CH=CH_2 \xrightarrow{BH_3} (CH_3CH_2CH_2CH_2)_3B \xrightarrow{\bar{O}H/H_2O_2}$$
1-Butene

$$\longrightarrow 3CH_3CH_2CH_2CH_2OH$$
1 - Butanol

(ii) Organoboranes on treatment with a carboxylic acids give the corresponding alkanes. This is a convenient method for the reduction of alkene and also alkynes to alkenes.

$$3CH_2CH=CH_2 \xrightarrow{(BH_3)_2} (CH_3CH_2CH_2)_3B \xrightarrow{3CH_3COOH} 3CH_3CH_2CH_3$$
1-Propene   Propane

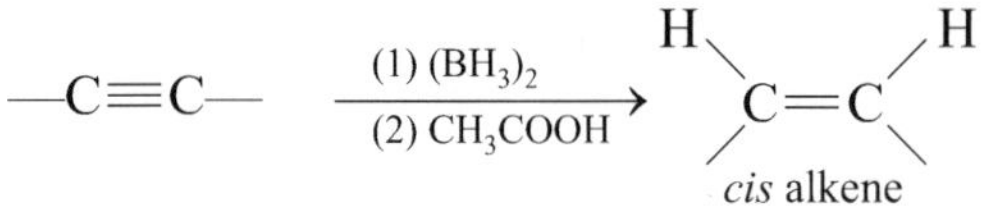

$$-C\equiv C- \quad \xrightarrow[\text{(2) CH}_3\text{COOH}]{\text{(1) (BH}_3)_2} \quad \overset{H}{\diagdown}C=C\overset{H}{\diagup}$$

cis alkene

(*iii*) The reaction of trialkylborone with alkaline AgNO$_3$ results in the synthesis of higher alkane. The reaction proceeds by a coupling reaction

$$2[(CH_3)_2CHCH_2)_2]B \xrightarrow{\text{AgNO}_3/\text{NaOH}} 3CH_3\overset{\overset{\displaystyle CH_3}{|}}{C}HCH_2CH_2\overset{\overset{\displaystyle CH_3}{|}}{C}HCH_3$$

2, 5-Dimethylhexane

2.6.1.3.2 Diels-Alder Reaction

The reaction involves reaction of a conjugated diene with unsaturated compounds called the dienophile (diene-lover) to yield an adduct. This reaction is a 1, 4-addition of an alkene to a conjugated diene and is called Diels-Alder Reaction.

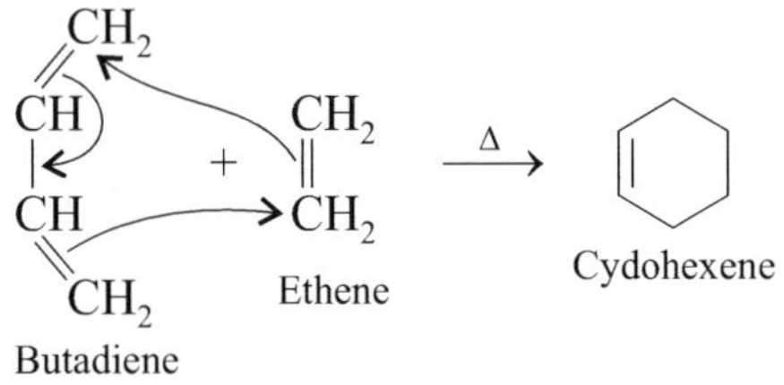

Butadiene   Ethene   Cydohexene

The reaction is of wide applicability as compounds containing multiple bonds other than carbon-carbon double bond can be used. In case, cyclic diene is used in the Diels-Alder reaction, bicyclic adducts are formed. Some typical examples of Diels-Alder reaction are given below:

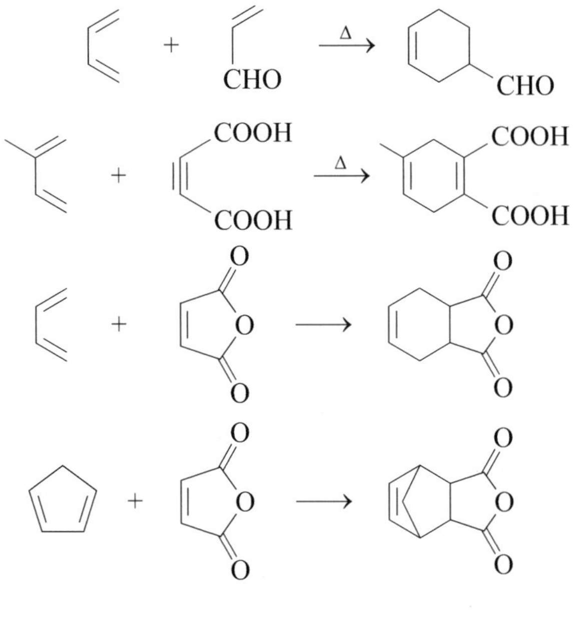

An important feature of Diels-Alder reaction is that it is stereospecific. During the reaction, the stereochemistry of the starting dienophilic is maintained. As an example maleic acid gives a *cis* product and fumaric acid gives a *trans* product.

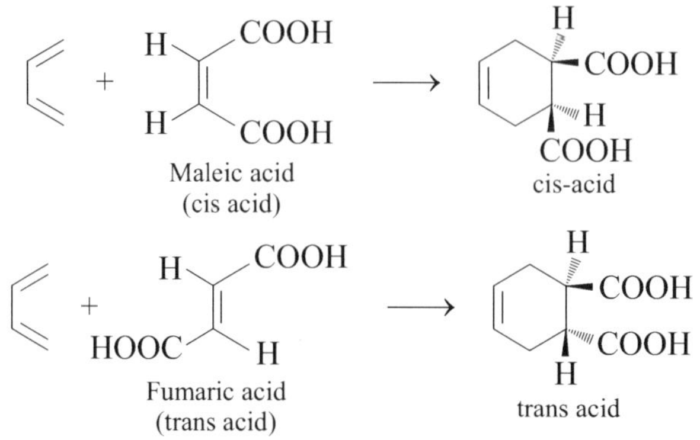

Maleic acid
(cis acid)

cis-acid

Fumaric acid
(trans acid)

trans acid

In Diels-Alder reaction, the dienes can react only when it adopts the *s-cis* form. Thus, an acyclic diene in *s-trans* conformation must rotate about the central C—C sigma bond to form the *s-cis* conformation for the reaction to proceed.

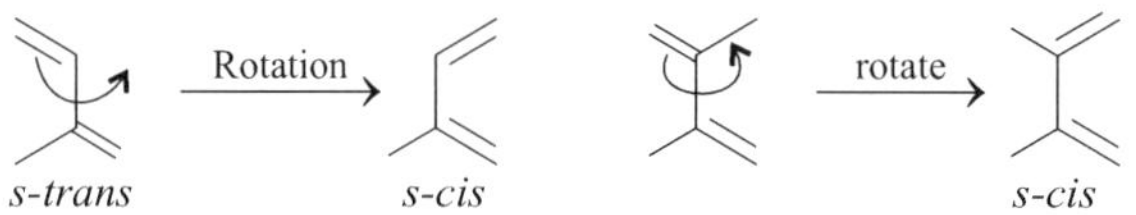

s-trans   *s-cis*   *s-cis*

Typical examples of reactive and unreactive cyclic dienes are given below:

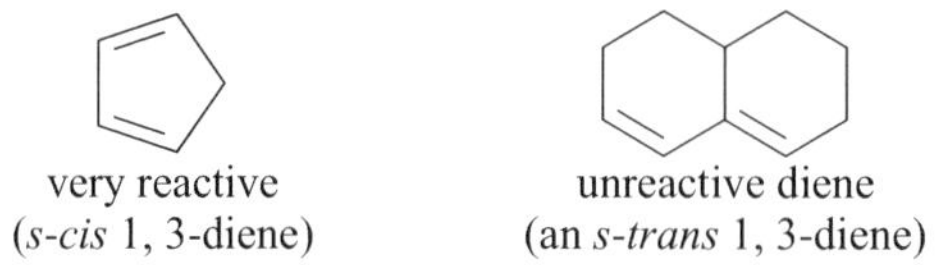

very reactive
(*s-cis* 1, 3-diene)

unreactive diene
(an *s-trans* 1, 3-diene)

2.6.1.3.3 Ozonolysis

Ozonolysis is a cleavage reaction in which the double bond is broken and the alkene is converted into two smaller molecules

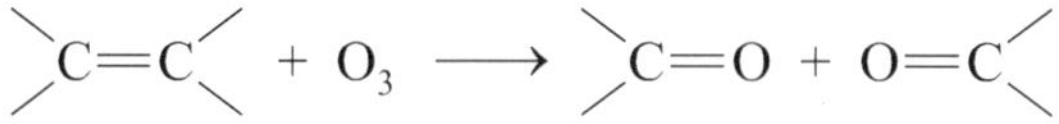

Ozonolysis, in fact, involves two separate reactions, *viz.,* oxidation of alkene or alkyne by ozone to yield an ozonide and reaction of the formed ozonide either by an oxidizing or reducing agent to yield cleavage products.

As an example the ozonide of 2-methyl-2-butene on reduction yields an aldehyde and a ketone. However oxidation of the ozonide gives a carboxylic acid and a Ketone.

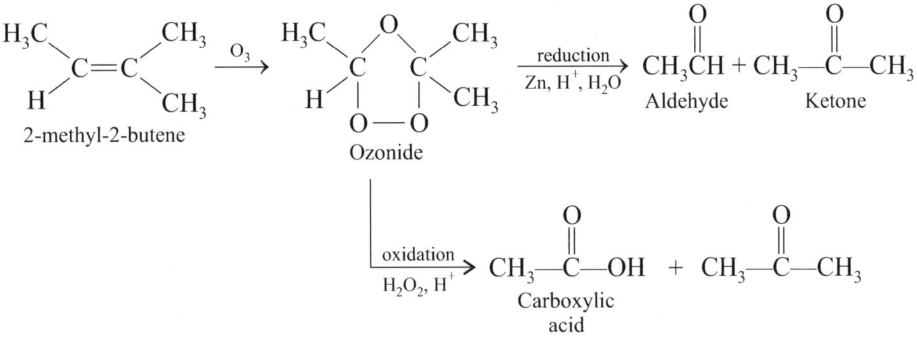

H₃C, CH₃ / C=C / H, CH₃ — 2-methyl-2-butene

$\xrightarrow{O_3}$

H₃C, O, CH₃ / C—C / H | | CH₃ / O—O — Ozonide

$\xrightarrow[\text{Zn, H}^+\text{, H}_2\text{O}]{\text{reduction}}$ CH₃CH + CH₃—C—CH₃

Aldehyde   Ketone

$\xrightarrow[\text{H}_2\text{O}_2\text{, H}^+]{\text{oxidation}}$ CH₃—C—OH + CH₃—C—CH₃

Carboxylic acid

The mechanism of the formation of the ozonide from the alkene is given below:

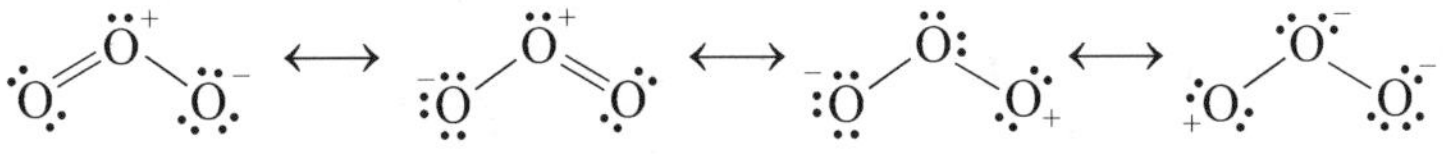

(Resonance forms of ozone)

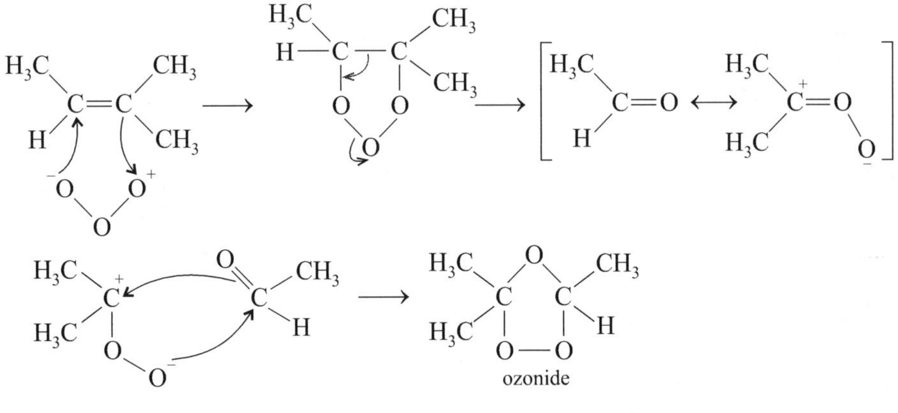

As seen, during ozonolysis, the first step is the conversion of 1, 3-dipolar addition of ozone to the double bond forms a molozonide. The molozonide, being unstable decomposes into fragments. These fragments on recombination in an alternative way yield the ozonide.

Ozonolysis is used to locate the position of the double bond in the starting alkene. An example is given below:

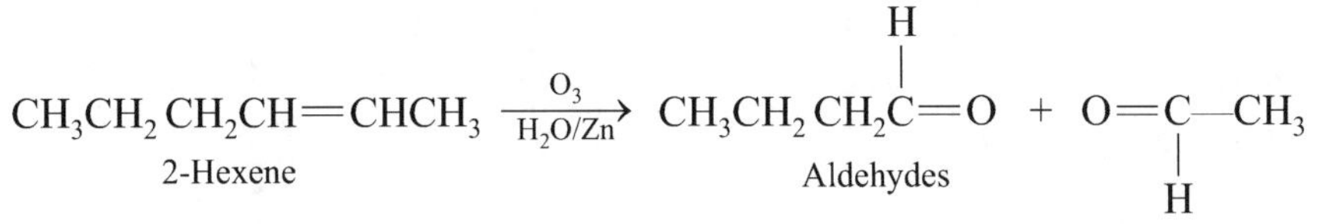

CH₃CH₂ CH₂CH=CHCH₃ $\xrightarrow[\text{H}_2\text{O/Zn}]{O_3}$ CH₃CH₂ CH₂C=O + O=C—CH₃

2-Hexene   Aldehydes

2.6.1.3.4 Hydroxylation

Hydroxylation of alkenes gives dihydroxy compounds known as diols. The hydroxylation of alkene can be carried out using a cold aqueous solution of $KMnO_4$ or $Os\,O_4$

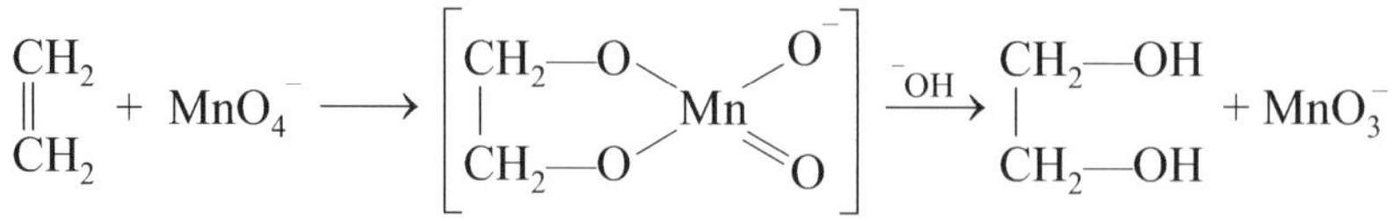

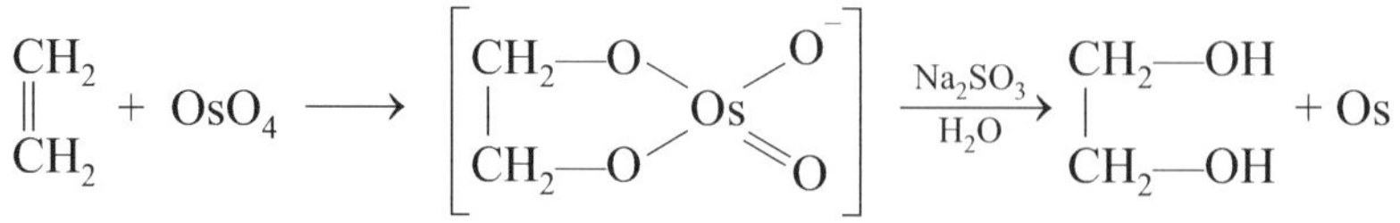

$$\underset{\text{CH}_2}{\overset{\text{CH}_2}{\|}} + \text{MnO}_4^- \longrightarrow \left[\begin{array}{c}\text{CH}_2-\text{O} \\ | \qquad\qquad \text{Mn} \\ \text{CH}_2-\text{O}\end{array}\overset{\text{O}^-}{\underset{\text{O}}{}}\right] \overset{\text{OH}^-}{\longrightarrow} \underset{\text{CH}_2-\text{OH}}{\overset{\text{CH}_2-\text{OH}}{|}} + \text{MnO}_3^-$$

$$\underset{\text{CH}_2}{\overset{\text{CH}_2}{\|}} + \text{OsO}_4 \longrightarrow \left[\begin{array}{c}\text{CH}_2-\text{O} \\ | \qquad\qquad \text{Os} \\ \text{CH}_2-\text{O}\end{array}\overset{\text{O}}{\underset{\text{O}}{}}\right] \overset{\text{Na}_2\text{SO}_3}{\underset{\text{H}_2\text{O}}{\longrightarrow}} \underset{\text{CH}_2-\text{OH}}{\overset{\text{CH}_2-\text{OH}}{|}} + \text{Os}$$

The oxidation of alkene (disappearance of pink colour) by cold aq. $KMnO_4$ is a well-known test (Baeyer test) for inferring the presence of a olefinic bond.

In both the above reactions, hydroxylation occurs with syn stereochemistry to yield *cis*-diol. However, *trans*-diols, as we have seen is obtained by the reaction of alkene with BH_3 followed by oxidation of the formed organoborane with alkaline H_2O_2. Trans hydroxylation of alkenes can also be effected by treatment of the alkene with an peracid and ring opening of the formed epoxide with hydrolytic cleavage.

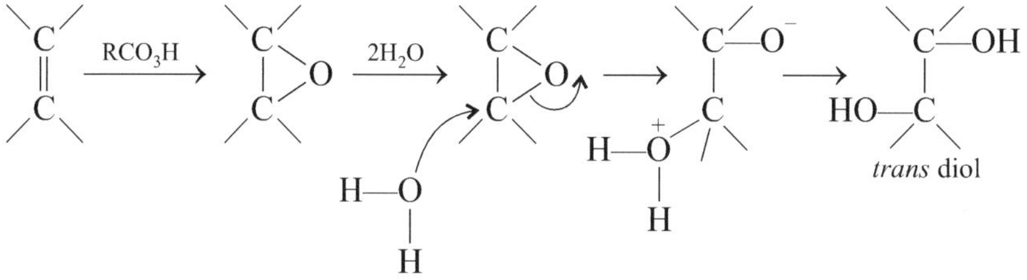

2.6.1.2 Nucleophilic Additions to Carbonyl Compounds

We have see that alkenes undergo electrophilic addition reactions. However aldehydes and ketones, which contain a carbonyl group undergo nucleophilic addition reactions.

The carbonyl carbon atom is sp^2 hybridised and forms three bonds (two C—H bonds and one C—O bond) and an unhybridised p-orbital is left on the carbon atom. Unlike carbon-carbon double bond, the carbon-oxygen double bond (in aldehydes and ketones) is polar. This is due to the higher electro negativity of oxygen compared to carbon. The π electrons of carbon-carbon double bond get shifted towards oxygen atom and the bond gets polarised.

Due to this there is imbalance in the π bond making the carbon atom electron deficient and as a result the carbonyl group as a whole has an electron withdrawing effect.

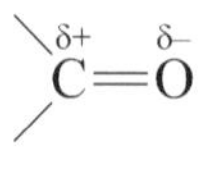

$$\overset{\delta+}{\underset{}{C}}=\overset{\delta-}{O}$$

The carbon carrying a partial positive charge can be attacked by nucleophilic reagent.

The ketones (containing two electron releasing alkyl groups) are less reactive than the aldehydes. Also chloroacetaldehyde ($ClCH_2CHO$) (which contain electron withdrawing chlorine atom) is more reactive than acetaldehyde. Similarly, nitro acetaldehyde (O_2NCH_2CHO) (which has stronger electron withdrawing character than chlorine) is more reactive than chloroacetaldehyde. Thus, the order of reactivity is:

$$CH_3CH_2{\rightarrow}CH < CH_3{\rightarrow}CH < Cl{\leftarrow}CH_2{-}CH < O_2N{\leftarrow}CH_2{-}CH$$

The aromatic aldehydes and ketones are less reactive than aliphatic aldehydes and ketones. This is due to the resonance interaction between the carbonyl group and the aromatic ring. The net result of this is a weakening of the positive charge on the carbonyl carbon atom *via.* dispersal of the charge within the ring.

The relative reactivity of aldehydes and ketones is also effected due to steric reasons. The presence of a bulky group in the vicinity of the carbonyl provides steric hinderence than the smaller hydrogen to the approaching nucleophile.

As already stated, aldehydes and ketones undergo nucleophilic addition reactions. These reactions take place in two step. In the first step, the nucleophile (:$\overline{Nu}$) attacks the electrophilic carbonyl. This results is the formation of a new bond to the nucleophile, the π bond breaks, moving an electron pair out on the oxygen atom. This results in the formation of an sp^3 hybridised intermediate. In the next step, protonation of the negatively charged oxygen atom by H_2O or other proton source forms the addition product.

Following are given some of the important nucleophilic addition reactions of carbonyl compounds.

2.6.1.2.1 Reaction with Hydrogen Cyanide

Hydrogen cyanide (hydrocyanic acid) adds to the carbonyl compounds in aqueous solution to give cyanohydrins (hydroxy nitriles)

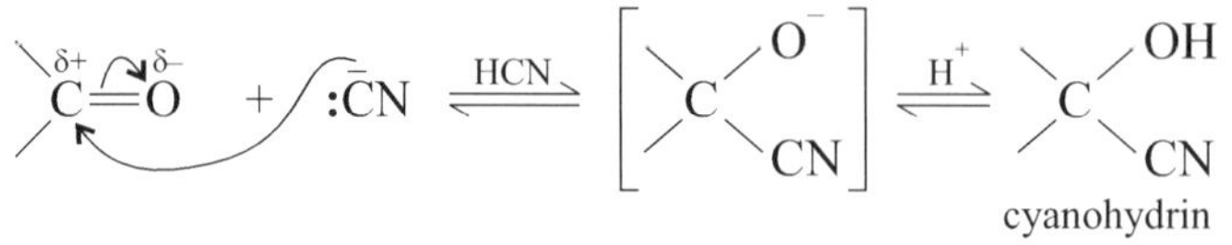

cyanohydrin

using benzaldehyde, mandelonitrile can be prepared

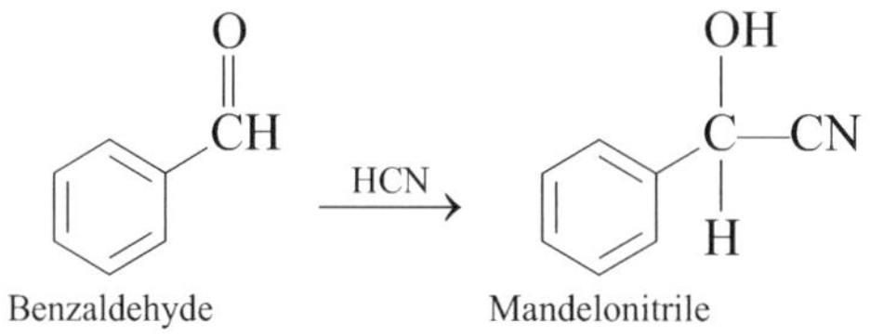

Benzaldehyde   Mandelonitrile

The cyanohydrins are useful intermediates for organic synthesis. Some example are

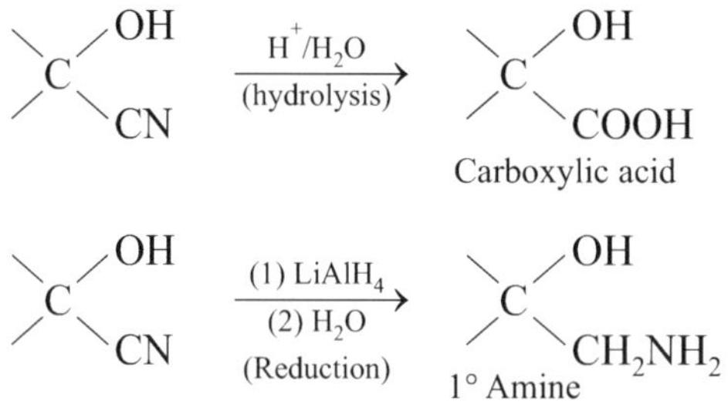

Carboxylic acid

1° Amine

2.6.1.2.2 Reaction with Sodium Hydrogen Sulphite

Sodium hydrogen sulphite ($NaHSO_3$) adds on to aldehydes, some ketones (mostly methyl ketones) and unhindered cyclic ketones to give crystalline hydrogen sulphite adduct.

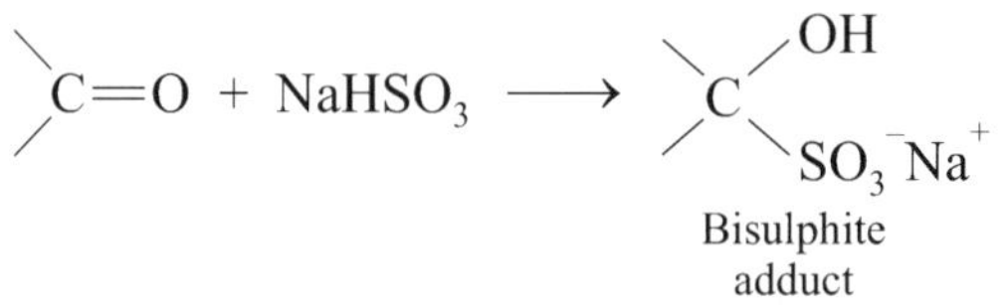

Bisulphite
adduct

Ketones with bulky groups do not add sodium hydrogen sulphite.

The bisulphite adduct on heating with dilute acid or aqueous Na_2CO_3 regenerate the starting carbonyl compound.

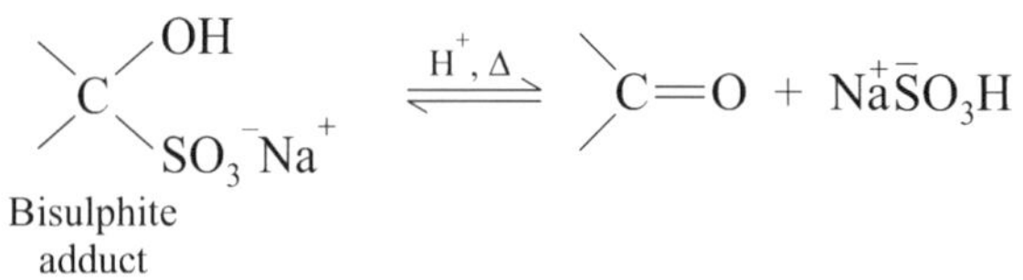

Bisulphite
adduct

This reaction is used for the separation and purification of carbonyl compounds from non-carbonyl compounds.

2.6.1.2.3 Reaction with Alcohols

Addition of one molecule of an alcohol to an aldehyde or ketone gives hemiacetal or hemiketal respectively. However, addition of two molecules of alcohol to an aldehyde or ketone with loss of water gives acetal or ketal respectively.

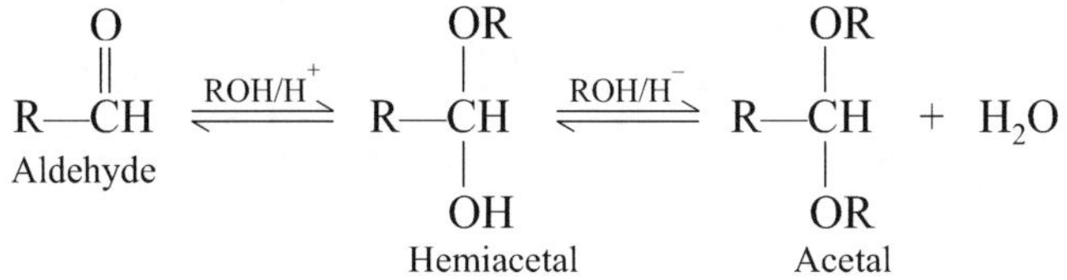

$$
\underset{\text{Aldehyde}}{\overset{\overset{\displaystyle O}{\|}}{R-CH}} \;\underset{}{\overset{ROH/H^+}{\rightleftharpoons}}\; \underset{\underset{OH}{|}}{\overset{\overset{\displaystyle OR}{|}}{R-CH}} \;\underset{}{\overset{ROH/H^-}{\rightleftharpoons}}\; \underset{\underset{OR}{|}}{\overset{\overset{\displaystyle OR}{|}}{R-CH}} + H_2O
$$

Hemiacetal   Acetal

The mechanism of acetal formation is given below:

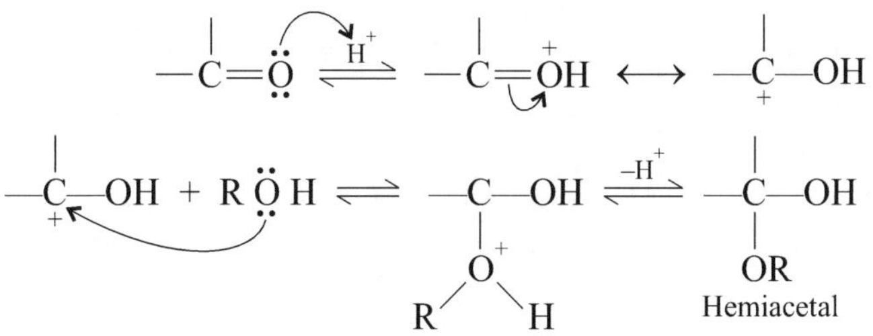

Mechanism for the formation of acetal from hemiacetal is given below:

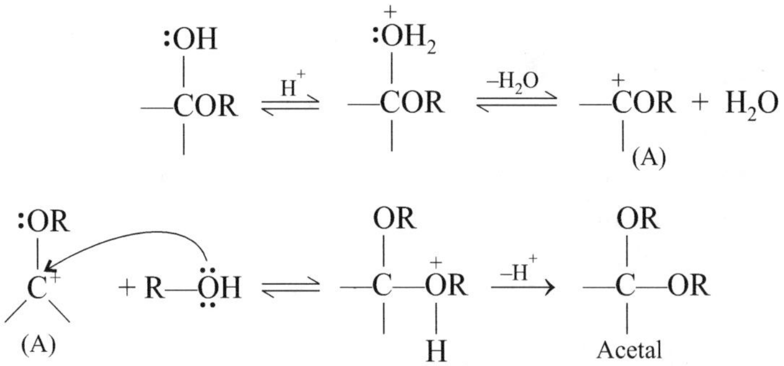

The acetal can be hydrolysed back into the parent carbonyl compound:

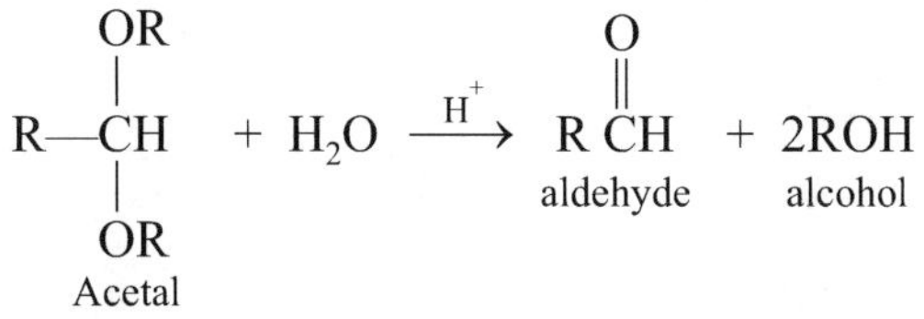

$$
\underset{\underset{OR}{|}}{\overset{\overset{\displaystyle OR}{|}}{R-CH}} + H_2O \;\xrightarrow{H^+}\; \underset{\text{aldehyde}}{\overset{\overset{\displaystyle O}{\|}}{R\,CH}} + \underset{\text{alcohol}}{2ROH}
$$

Acetal

Formation of acetal and hydrolysis back to aldehyde is useful in synthetic organic chemistry. As an example, the conversion of acrylaldehyde $\left(\underset{}{\overset{\overset{\displaystyle O}{\|}}{CH_2\!\!=\!\!CHCH}}\right)$ into

the corresponding diol, 2, 3-dihydroxypropanal $\left(\underset{}{\overset{\overset{\displaystyle O\quad O}{\|\;\;\;\|}}{HOCH_2CH\!-\!CH}}\right)$ cannot be done directly. For this conversion the CHO has to be protected by treatment with ethylene glycol in presence of acid catalysis and the protected compound treated will dilute $KMnO_4$/alkali to give the diol. Final deprotection regenerates the aldehyde group

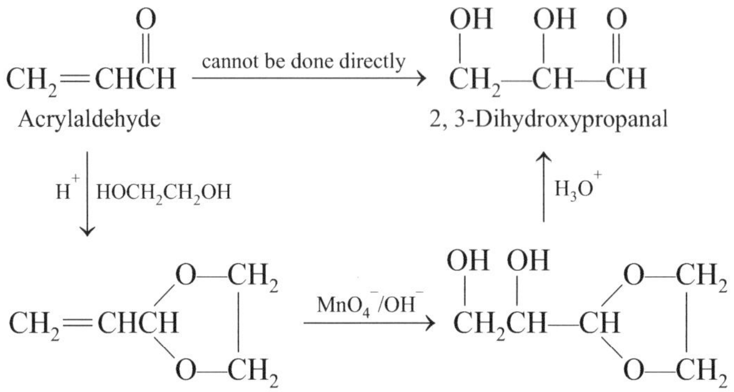

$$CH_2=CHCH\overset{O}{\overset{||}{}} \xrightarrow{\text{cannot be done directly}} CH_2-CH-CH\overset{OH\ OH\ O}{\overset{|\ \ |\ \ ||}{}}$$

Acrylaldehyde   2, 3-Dihydroxypropanal

$$H^+ \Big| HOCH_2CH_2OH \qquad\qquad\qquad\qquad \Big\uparrow H_3O^+$$

$$CH_2=CHCH\overset{O-CH_2}{\underset{O-CH_2}{\diagdown \diagup}} \xrightarrow{MnO_4^-/OH^-} CH_2CH-CH\overset{OH\ OH}{\overset{|\ \ |}{}}\overset{O-CH_2}{\underset{O-CH_2}{\diagdown \diagup}}$$

Ketones do not normally give acetals, but they give Ketal by treatment with ethylene glycol in presence of acid catalyst

2.6.1.2.4 Reaction with Amines

Primary amines, RNH_2 react with carbonyl compounds to give an imine (a compound containing C=N group)

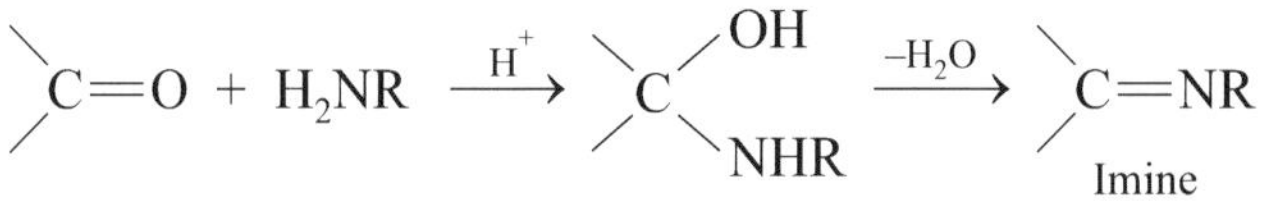

$$\overset{\diagdown}{\underset{\diagup}{}}C=O + H_2NR \xrightarrow{H^+} \overset{\diagdown}{\underset{\diagup}{}}C\overset{OH}{\underset{NHR}{\diagup}} \xrightarrow{-H_2O} \overset{\diagdown}{\underset{\diagup}{}}C=NR$$

Imine

The amines are also known as Schiff's bases. Some examples of Schiff's base formation are given below:

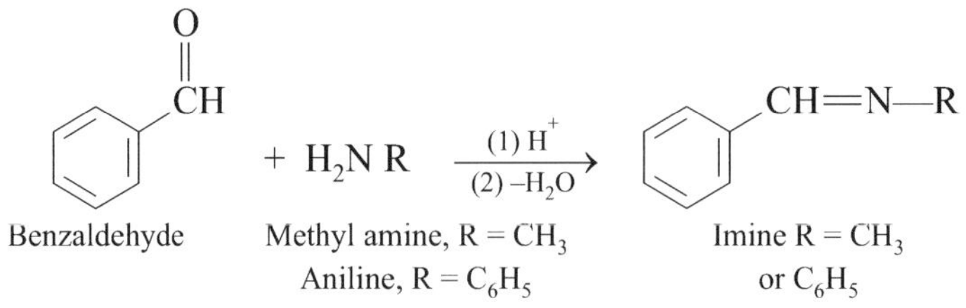

Benzaldehyde   + H$_2$N R $\xrightarrow[(2)\,-H_2O]{(1)\,H^+}$ Imine R = CH$_3$
                    Methyl amine, R = CH$_3$ or C$_6$H$_5$
                    Aniline, R = C$_6$H$_5$

The formation of Schiff's base is given below:

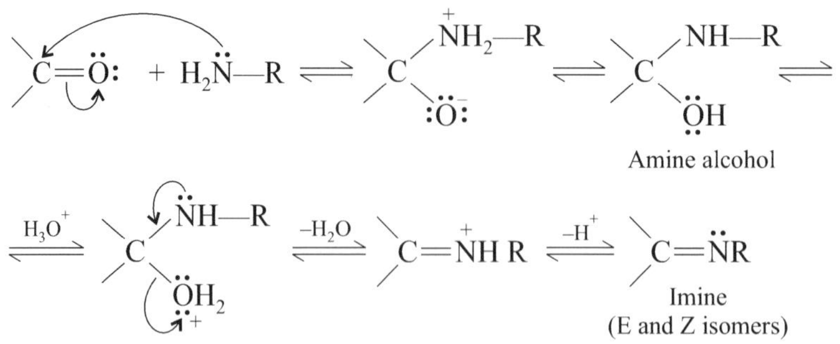

$$\overset{\diagdown}{\underset{\diagup}{}}C=\overset{..}{\underset{..}{O}} + H_2\overset{..}{N}-R \rightleftharpoons \overset{\diagdown}{\underset{\diagup}{}}C\overset{\overset{+}{N}H_2-R}{\underset{\overset{..}{\underset{..}{O}}:}{}} \rightleftharpoons \overset{\diagdown}{\underset{\diagup}{}}C\overset{NH-R}{\underset{\overset{..}{\underset{..}{O}}H}{}} \rightleftharpoons$$

Amine alcohol

$$\overset{H_3O^+}{\rightleftharpoons} \overset{\diagdown}{\underset{\diagup}{}}C\overset{\overset{..}{N}H-R}{\underset{\overset{..}{\underset{+}{O}}H_2}{}} \overset{-H_2O}{\rightleftharpoons} \overset{\diagdown}{\underset{\diagup}{}}C=\overset{+}{N}H R \overset{-H^+}{\rightleftharpoons} \overset{\diagdown}{\underset{\diagup}{}}C=\overset{..}{N}R$$

Imine
(E and Z isomers)

Following table gives the reaction of carbonyl compounds with various amine derivatives

R in RNH$_2$	Amine derivatives (Reagent) Name	Product	Name
—H	Ammonia NH$_3$	$\diagdown$C=NH$\diagup$ (E) and (Z) isomers	Imine (unstable)
—R	Amine R—NH$_2$	$\diagdown$C=NR$\diagup$	Substituted imine (Schiff's base)
—OH	Hydroxylamine NH$_2$OH	$\diagdown$C=N—OH$\diagup$ (E) and (Z) isomers	Oxime
—NH$_2$	Hydrazine NH$_2$—NH$_2$	$\diagdown$C=N—NH$_2$$\diagup$ (E) and (Z) isomers	Hydrazones
—NHC$_6$H$_5$	Phenyl hydrazine C$_6$H$_5$NHNH$_2$	$\diagdown$C=N—C$_6$H$_5$$\diagup$	Phenyl hydrazone
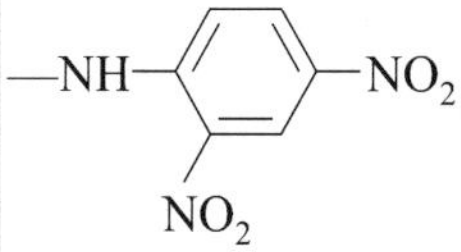 —NH—⬡—NO$_2$ (with NO$_2$)	2, 4-Dinitro Phenyl hydrazine	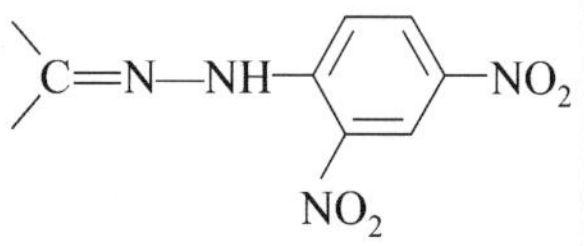 $\diagdown$C=N—NH—⬡—NO$_2$$\diagup$ (with NO$_2$)	2, 4-Dinitro phenyl hydrazone
—HN—C(=O)—NH$_2$	Semicarbazide H$_2$N NH CONH$_2$	$\diagdown$C=N NH CONH$_2$$\diagup$	Semicarbazone

Secondary amines react with aldehydes and ketones having an α-hydrogen yield iminium ions, which undergo further reaction to give enamines (vinylamines)

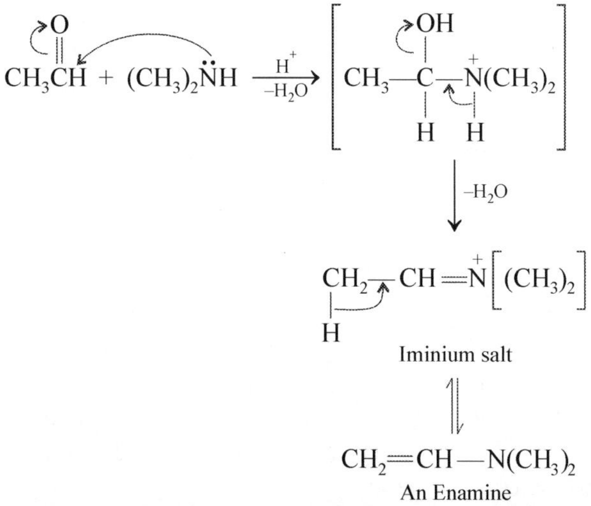

$$CH_3\overset{\curvearrowleft}{C}\overset{O}{\overset{\parallel}{H}} + (CH_3)_2\ddot{N}H \xrightarrow[-H_2O]{H^+} \left[CH_3-\overset{OH}{\underset{\underset{H}{|}}{C}}\overset{\curvearrowleft}{-}\overset{+}{N}(CH_3)_2 \right]$$

$$\downarrow{-H_2O}$$

$$\underset{H}{\overset{|}{CH_2}}\overset{\curvearrowright}{-}CH=\overset{+}{N}\left[(CH_3)_2\right]$$

Iminium salt

$$\Updownarrow$$

$$CH_2=CH-N(CH_3)_2$$

An Enamine

Like imine formation, enamine formation is also reversible and can be converted back to the corresponding carbonyl compounds.

2.6.1.2.5 Reaction with Grignard Reagents

The Grignard reagent is best prepared by the reaction of alkyl halide in ether by the process of sonication

$$R-X + Mg \xrightarrow[\text{))))}]{\text{ether}} RMgX$$
$$(90\%)$$

The Grignard reagents react with aldehydes or Ketones to give alcohols.

$$\overset{|}{\underset{|}{C}}=O + RMgX \longrightarrow \overset{|}{\underset{|}{C}} \overset{OMgX}{\underset{R}{}} \xrightarrow{-H_2O} \overset{|}{\underset{|}{C}} \overset{OH}{\underset{R}{}}$$

This is a convenient method for the preparation of alcohols.

Thus, formaldehyde on reaction with CH_3 Mg Br followed by hydrolysis given primary alcohols

$$\underset{\text{Formaldehyde}}{\overset{O}{\overset{||}{H\,CH}}} + CH_3MgBr \longrightarrow \underset{H}{\overset{H}{\diagdown}}C\underset{CH_3}{\overset{OMgX}{\diagup}} \xrightarrow{H_2O/H^+} \underset{\text{Ethanol}}{CH_3CH_2OH}$$

Other aldehydes (except formaldehyde) yield secondary alcohol

$$CH_3\,CH_2\,\overset{O}{\overset{||}{CH}} \xrightarrow[(2)\,H_2O,\,H^+]{(1)\,CH_3MgBr} CH_3\,CH_2\underset{\underset{CH_3}{|}}{CH}\,OH$$
$$\text{Propanal} \qquad\qquad\qquad \text{2-Butanol}$$

Ketones on reaction with grignard reagent yield tertiary alcohols

$$\underset{H_5C_2}{\overset{H_5C_2}{\diagdown}}C=O \xrightarrow[(2)\,H_2O,\,H^+]{(1)\,CH_3MgBr} C_2H_5\underset{\underset{CH_3}{|}}{\overset{\overset{C_2H_5}{|}}{C}}OH$$
$$\text{3-petanone} \qquad\qquad\qquad \text{3-methyl-3-pentanol}$$

Mechanism

It is believed that the Grignard reaction occurs in the following ways:

(i) The Grignard reagent being strongly nucleophilic uses its electron pair to form a bond to the carbon atom of the carbonyl group. One electron pair of the carbonyl group shifts out to the oxygen. This reaction is a nucleophilic addition to the carbonyl group and it results in the formation of an alkoxide ion associated with Mg^{2+} and X^-; the adduct is called halomagnesium alkoxide.

(*ii*) The next step is the protonation of the alkoxide ion by the addition of aqueous hydrogen halide (HX). This results in the formation of the alcohol and MgX$_2$.

The various step are represented as shown below:

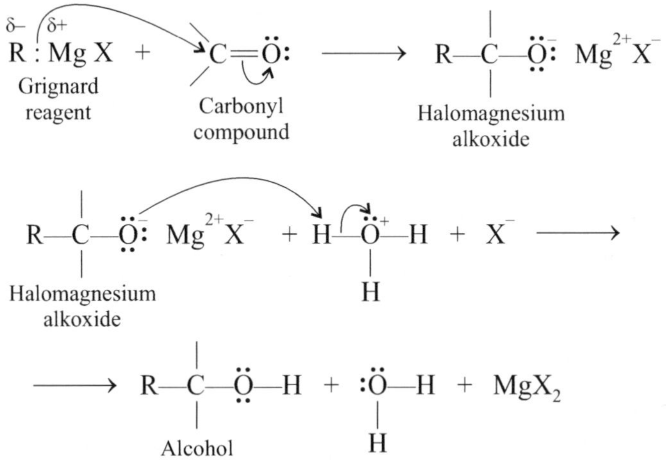

$$\overset{\delta-}{R} : \overset{\delta+}{Mg} X \;+\; {>}C{=}\ddot{O}: \longrightarrow R{-}\overset{|}{\underset{|}{C}}{-}\ddot{\underset{..}{O}}: \; Mg^{2+}X^-$$

Grignard reagent   Carbonyl compound   Halomagnesium alkoxide

$$R{-}\overset{|}{\underset{|}{C}}{-}\ddot{\underset{..}{O}}: \; Mg^{2+}X^- \;+\; H{-}\overset{H}{\underset{|}{\overset{+}{\ddot{O}}}}{-}H \;+\; X^- \longrightarrow$$

Halomagnesium alkoxide

$$\longrightarrow R{-}\overset{|}{\underset{|}{C}}{-}\ddot{\underset{..}{O}}{-}H \;+\; :\overset{H}{\underset{|}{\ddot{O}}}{-}H \;+\; MgX_2$$

Alcohol

2.6.1.2.6 Wittig Reaction

The reaction of carbonyl compounds (aldehydes or ketones) with phosphorous ylides (or phosphorane) (commonly known as Wittig reagent) yield alkenes and triphenylphosphine oxide.

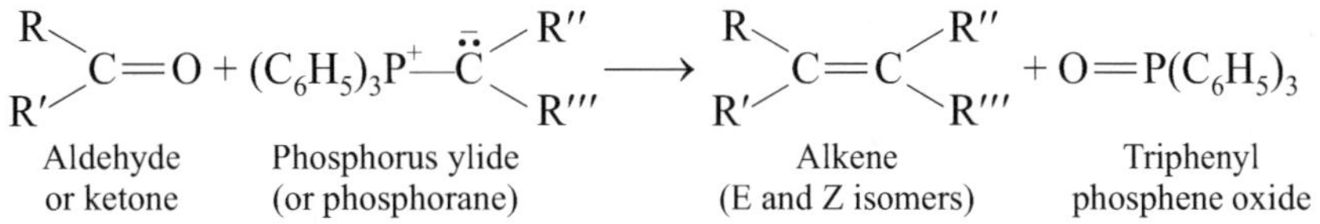

$$\overset{R}{\underset{R'}{>}}C{=}O + (C_6H_5)_3\overset{+}{P}{-}\overset{..}{\underset{}{C}}\overset{R''}{\underset{R'''}{<}} \longrightarrow \overset{R}{\underset{R'}{>}}C{=}C\overset{R''}{\underset{R'''}{<}} + O{=}P(C_6H_5)_3$$

Aldehyde or ketone   Phosphorus ylide (or phosphorane)   Alkene (E and Z isomers)   Triphenyl phosphene oxide

This reaction is known as the Wittig reaction and is a very convenient method for the synthesis of alkenes (a mixture of (E) and (Z) isomers result). In this reaction, there is absolutely no ambiguity as to the location of the double bond in the product, in contrast to E1 eliminations, which may yield multiple alkene products by rearrangement to more stable carbocation intermediate and both E1 and E2 elimination reactions may occur; this produces multiple products when different β-hydrogens are available for removal.

Witting reaction was discovered by George Wittig (and hence the name) in 1954 and was awarded Nobel Prize in Chemistry in 1979 due to its tremendous synthetic potentialities, being a valuable method for synthesising alkenes.

The Phosphorous Ylides

The phosphorus ylides (commonly known as Wittig reagent) are obtained by the reaction of an alkyl halide with triphenylphosphine. The formed phosphonium salt is treated with a strong base like sodium hydride or phenyl lithium to give phosphorus ylide. These phosphorus ylides carry a positive and negative charge on adjacent atoms and can be represented by doubly bonded species called phosphoranes.

$$Ph_3P: + \overset{\overset{\displaystyle R'}{\displaystyle |}}{CH_2}X \longrightarrow Ph_3\overset{+}{P}-CH_2R' \; X^- \xrightarrow{\text{Base}}$$

Triphenyl phosphene   Alkyl halide   Phosphonium salt

$$\longrightarrow [Ph_3\overset{+}{P}-\overset{-}{C}HR' \longleftrightarrow Ph_3P=CHR']$$

Ylide   Phosphorane

In the above procedure, the first step is a nucleophilic substitution reaction, triphenylphosphine being an excellent nucleophile and a weak base readily reacts with 1° and 2° alkyl halides by an S_N2 mechanism to displace a halide ion from the alkyl halide to give an alkyl triphenyl phosphonium salt. In the second step (which is an acid-base reaction), a strong base removes a proton from the carbon atom that is attached to phosphorus to give the ylide.

Mechanism

The mechanism of Wittig reaction has been the subject of considerable study. It was earlier suggested that the ylide, acting as a carbanion attacks the carbonyl carbon of the aldehyde or ketone to form an unstable intermediate with separated charge called betaine. In the subsequent step, the betaine being unstable gets converted into a four-membered cyclic system called an oxaphosphetane, which spontaneously loose a molecule of triphenyl phosphine oxide to form an alkene. Subsequent studies have shown that betanine is not an intermediate and that the oxaphosphetane is formed directly by a cycloaddition reaction. The driving force for the Wittig reaction is the formation of very strong phosphorus-oxygen bond in triphenylphosphine oxide.

A typical example is given below:

$$Ph_3P + CH_3Br \xrightarrow{C_6H_5Li} Ph_3\overset{+}{P}-CH_3Br^- \xrightarrow{C_6H_5Li}$$

Methyl triphenyl
Phosphonium bromide
89%

$$\longrightarrow Ph_3\overset{+}{P}-\overset{-}{C}H_2 \longleftrightarrow Ph_3P=CH_2 + C_6H_6 + LiBr$$

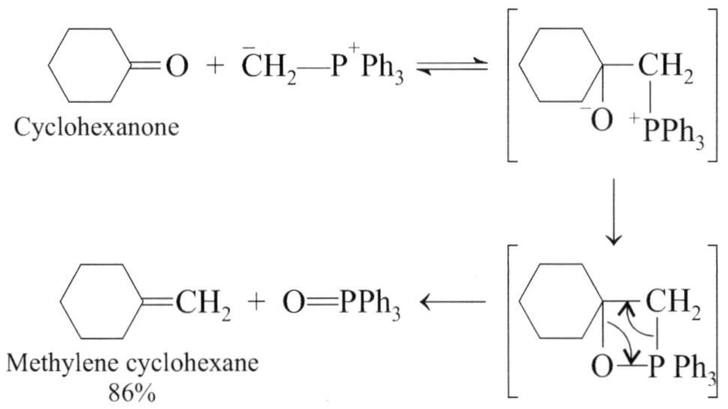

Cyclohexanone

Methylene cyclohexane
86%

Though Wittig synthesis appears to be complicated, in fact, these are easy to carry out as one pot reaction.

2.6.1.2.7 Aldol Condensation

For details see section 2.4.3.2

2.6.1.2.8 Perkin Condensation

It involves the reaction of aromatic aldehyde with anhydride of an aliphatic acid (having at least two alpha hydrogen atoms) in presence of the salt of the same acid (which acts as a catalyst) to yield an α, β-unsaturated acid. As an example, benzaldehyde on heating with acetic anhydride in presence of anhydrous sodium acetate gives cinnamic acid.

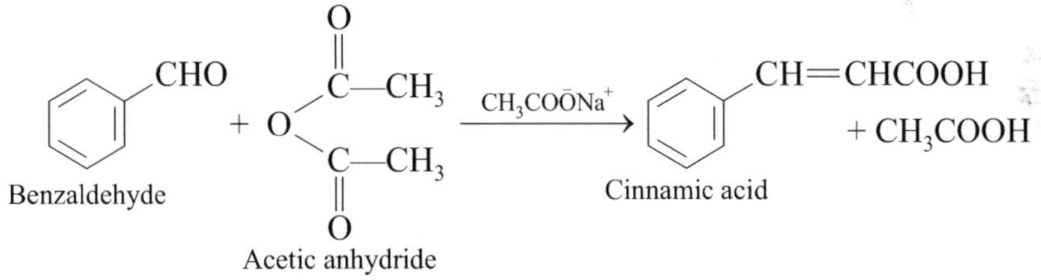

Benzaldehyde

Acetic anhydride

Cinnamic acid

The mechanism of the reaction is given below:

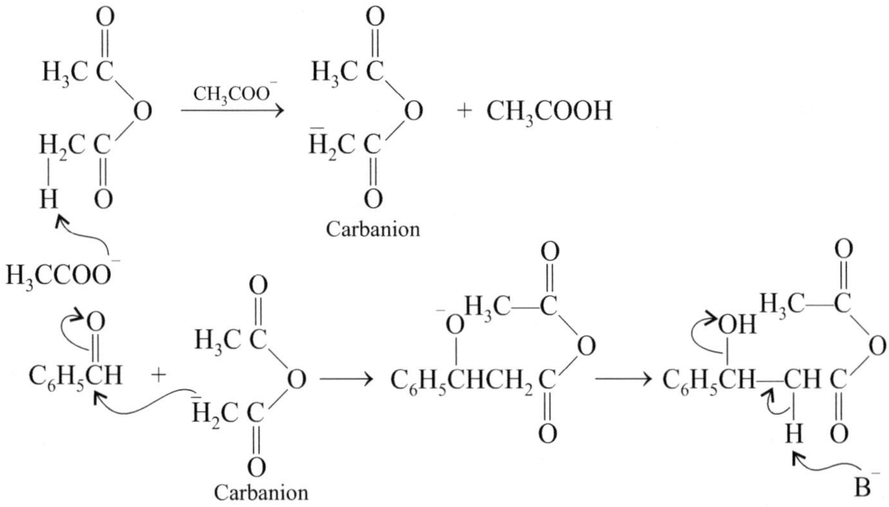

Carbanion

Carbanion

B

$$\longrightarrow C_6H_5CH=CH-C \overset{H_3C-C\overset{O}{\underset{O}{\parallel}}}{\underset{\parallel}{O}} \xrightarrow{\text{hydrolysis}} \underset{\text{cinnamic acid}}{C_6H_5CH=CH-COOH} + CH_3COOH$$

2.6.1.2.9 Claisen Condensation

It involves the reaction of esters having α-hydrogen with a base like sodium ethoxide, self condensation of esters occur resulting in the formation of a β-keto ester.

$$\underset{\text{Ethyl acetate}}{2CH_3\overset{O}{\overset{\parallel}{C}}OC_2H_5} \xrightarrow{\overset{+}{N}a\bar{O}Et} \underset{\text{Ethyl acetoacetate}}{CH_3\overset{O}{\overset{\parallel}{C}}CH_2\overset{O}{\overset{\parallel}{C}}OC_2H_5} + C_2H_5OH$$

The mechanism of the reaction involves following steps:

$$\underset{\underset{\bar{O}C_2H_5}{\overset{\displaystyle H}{\uparrow}}}{CH_2\overset{O}{\overset{\parallel}{C}}OC_2H_5} \rightleftharpoons \left[H_2\bar{C}-\overset{O}{\overset{\parallel}{C}}OC_2H_5 \longleftrightarrow H_2C=\overset{\bar{O}}{\overset{\displaystyle}{C}} OC_2H_5 \right]$$
$$\text{Ethyl enolate ion}$$

$$\underset{\bar{O}C_2H_5}{CH_3\overset{O}{\overset{\displaystyle}{C}}} + CH_2\overset{O}{\overset{\parallel}{C}}OC_2H_5 \rightleftharpoons CH_3\overset{\bar{O}}{\underset{\underset{OC_2H_5}{\displaystyle}}{C}}-CH_2\overset{O}{\overset{\parallel}{C}}-OC_2H_5 + \bar{O}C_2H_5$$

$$\downarrow -\bar{O}C_2H_5$$

$$\underset{\text{ethyl acetoacetate}}{CH_3\overset{O}{\overset{\parallel}{C}}-CH_2\overset{O}{\overset{\parallel}{C}}-OC_2H_5}$$

The claisen condensation between two different esters, both having an α-hydrogen is known as **crossed claisen condensation.** This reaction however is not of synthetic utility due to formation of a mixture of products

| Methyl benzoate | Acetone | 1-Phenyl-1, 3-butanedione |

$$\underset{\text{Methyl benzoate}}{\overset{O}{\overset{\parallel}{C}}OCH_3} + \underset{\text{Acetone}}{CH_3\overset{O}{\overset{\parallel}{C}}CH_3} \xrightarrow[\text{(2) H}^+]{\text{(1) NaOCH}_3} \underset{\text{1-Phenyl-1, 3-butanedione}}{\overset{O}{\overset{\parallel}{C}}CH_2\overset{O}{\overset{\parallel}{C}}CH_3}$$

2.6.1.2.10 Knoevenagel Condensation

It involves base catalysed condensation between an aldehyde or ketone with a compound having an active methylene group (*e.g.*, malonic ester)

$$CH_3CHO \quad + \quad CH_2(COOH)_2 \xrightarrow{base} CH_3CH=C(COOH)_2$$

Acetaldehyde · · · · · malonic acid

$$\downarrow \Delta-CO_2$$

$$CH_3CH=CHCOOH$$

crotonic acid

The base used in the above condensation is a weak base like ammonia or amine (primary or secondary) However, when condensation is carried out in presence of pyridine as base, decarboxylation occurs during the condensation. This is know as **Doebner modification**

$$CH_3CHO \quad + \quad CH_2(COOH)_2 \xrightarrow{base} C_6H_5CH=C(CO_2Et)_2$$

Benzaldehyde · · · · · Ethyl malonalte

$$H_3O^+ \downarrow \text{hydrolysis}$$

$$C_6H_5CH=CHCOOH \xleftarrow[-CO_2]{\Delta} C_6H_5CH=C(COOH)_2$$

Cinnamic acid

The Knoevenagel reaction is more useful with aromatic aldehydes, since with aliphatic aldehydes, the product obtained undergoes **Michael condensation**. As an example, tetraethyl propane-1,1,3,3-tetracarboxylate is obtained by Knoevenagel condensation of formaldehyde with diethyl malonate in the presence of diethylamine, followed by Michael addition reaction to yield the final product.

$$O=CH_2 \quad + \quad \bar{C}H(CO_2Et)_2 \underset{}{\overset{Et_2NH}{\rightleftharpoons}} HOCH_2-CH(CO_2Et)_2$$

Formaldehyde

$$\downarrow -H_2O$$

$$(EtO_2C)_2CHCH_2CH(CO_2Et)_2 \xleftarrow[\text{Michael addn.}]{\bar{C}H(CO_2Et)_2} CH_2=C(COOEt)_2$$

Adduct
Tetraethyl propane-
1,1,3,3-tetracarboxylate

The effectiveness of various activating groups in the active methylene compounds is found to be in the order

$$NO_2 > CN > COCH_3 > COC_6H_5 > COOC_2H_5$$

Ketones do not undergo Knoevenagel condensation with malonic ester, but can react with more active cyanoacetic acid or its ester. For example, acetone forms isopropylidene cyanoacetic ester when condensed with ethyl cyanoacetate.

$$(CH_3)_2C=O \quad + \quad CH_2 \overset{CN}{\underset{CO_2Et}{<}} \longrightarrow (CH_3)_2C=C \overset{CN}{\underset{CO_2Et}{<}}$$

Acetone · · · · · Ethyl cyanoacetate · · · · · Isopropylidene
cyanoacetic ester

The Knoevenagel reactions is reversible and the equilibrium normally lies towards left giving low yields. The yield can be improved by carrying out the reaction in benzene and removal of the formed water by azeotropic distillation using a Dean stark apparatus. This is referred as **Cope-Knoevenagel reaction.**

MECHANISM

The base removes a proton from the active methylene compound to give a carbanion (which is resonance stabilized). The carbanion then attacks the carbonyl carbon of the aldehyde or ketone. Subsequent protonation of the anion followed by dehydration yields the product.

$$CH_2(CO_2Et)_2 \xrightarrow{\text{:B}} {}^-CH(CO_2Et)_2$$

$$R—CH=C(CO_2Et)_2 \xleftarrow{\text{—H}_2O} R—\underset{\underset{H}{|}}{\overset{\overset{OH}{|}}{C}}—CH(CO_2Et)_2$$

Ketones or aldehydes also react with a succinic ester in presence of sodium hydride to give the corresponding condensation product.

This condensation is known as **Knoevenagel-Stobbe condensation.**

2.6.1.2.11 Cannizzaro Reaction

Aldehydes without α-hydrogen (S) on treatment with concentrated aqueous alkali undergo self-oxidation and reduction to give an alcohol and the salt of the corresponding carboxylic acid

$$\underset{\text{Benzaldehyde}}{C_6H_5CHO} + C_6H_5CHO \xrightarrow{\text{KOH}} \underset{\text{Benzyl alcohol}}{C_6H_5CH_2OH} + \underset{\text{Pot. benzoate}}{C_6H_5COOK}$$

This disproportion or self-oxidation and reduction of aromatic aldehydes, devoid of α-hydrogen is known as cannizzaro reaction. The reaction best proceeds with aromatic aldehydes without α-hydrogen. However some aliphatic aldehydes like formaldehyde and dimethyl acetaldehyde (which do not have α-hydrogen) also undergo cannizzaro reaction.

$$HCHO + NaOH \xrightarrow{\Delta} CH_3OH + HCOONa$$

Formaldehyde · · · · · · · · · · · · · · Methyl alcohol · · Sod. formate

$$2(CH_3)_2CHCHO + NaOH \xrightarrow{\Delta} (CH_3)_2CHCH_2OH + (CH_3)_2CHCOONa$$

Dimethyl acetaldehyde · · · · · · · · · · 2-Methyl-1-propanol · · Sod. 2-methyl-1-propionate

(Scheme-2)

MECHANISM

Cannizzaro reaction involves transfer of a hydrogen atom from one molecule of the aldehyde to another. This has been established by using deuterated benzaldehyde instead of benzaldehyde.

$$2C_6H_5CDO + OH^- \xrightarrow{H_2O} C_6H_5CD_2OH + C_6H_5\overset{\overset{\displaystyle O}{\|}}{C}-O^-$$

It is found that the benzyl alcohol formed is exclusively deuterated. So, the possibility of exchange of hydrogen with hydrogen atoms in the solvent is ruled out. Thus, a transfer of hydrogen (or deuterium) takes place between the carbonyl atoms of the two aldehyde molecules.

Cannizzario reaction proceeds by formation of an anion (by reaction with base) which may transfer a hydride ion intermolecularity to the carbonyl of another aldehyde molecule forming the carboxylic acid and the alkoxide ion. Final step is the shifting of a proton from the acid to the alcohol.

If the Cannizzaro reaction is performed in D_2O, the alcohol formed has no deuterium establishing that the mechanism involves a direct transfer of hydrogen from one molecule of aldehyde to another as already depicted.

The Cannizzaro reaction under heterogenous conditions catalysed by barium hydroxide is considerably accelerated by sonication. The yields are 100 percent after 10 min. whereas no reaction is observed during this period without ultrasound.

$$C_6H_5CHO \xrightarrow[\text{)))), 10 min}]{\text{Ba(OH)}_2\text{EtOH}} C_6H_5CH_2OH + C_6H_5COOH$$

Benzyl alcohol is obtained by the crossed Cannizzaro reaction of benzaldehyde and formaldehyde.

$$\underset{\text{Benzaldehyde}}{C_6H_6CHO} + CH_2O \xrightarrow{\text{30\% NaOH}} \underset{\text{Benzyl alcohol}}{C_6H_5CH_2OH} + \underset{\text{Forminc acid}}{HCO_2H}$$

2.6.1.2.12 Michael Addition

For details see section 2.4.3.2.

2.6.2 Elimination reactions

Elimination reactions are the reverse of addition reactions. Treatment of an alkyl halide with a base results in elimination.

$$RCH_2CH_2X \xrightarrow{\text{}^-OH} RCH{=}CH_2 + X^-$$

In elimination reaction, there are loss of two atoms or a group from a molecule. We come across many types of elimination reactions. These are 1, 1-elimination, 1, 2-elimination and 1, 3-elimination etc.

In 1, 1-elimination, both the atoms or groups are lost from the same carbon atom. For a discussion see section 2.4.1.1. This method finds use in the generation of dichlorocarbene from chloroform (see section 2.4.3.4). The reaction is called α-elimination.

In 1, 2-elimination, both the atoms or groups are lost from two adjacent carbon atoms resulting in the formation of a double or triple bond. The reaction is called β-elimination.

In 1, 3-elimination, both the atoms or groups are lost from carbons 1 and 3 resulting in the formation of a 3-membered ring.

In 1, 4-elimination, both the atoms or groups are lost from carbons 1 and 4.

The present discussion is restricted to 1, 2-eliminations only.

The 1, 2-eliminations are of two types, *viz.*, Bimolecular elimination reactions (E2) and unimolecular elimination reactions (E1).

2.6.2.1 Bimolecular Elimination Reactions (E2)

In these elimination reactions, the rate of elimination depends on the concentration of the both the substrate and the nucleophile and the reaction is of second order. It is represented as E2. The E2 reaction is a one-step process. In this, abstraction of the proton from the β carbon atom and the expulsion of the leaving group (*i.e.*, halide ion) from the α carbon atom occur simultaneously. The mechanism of this reaction is given as follows:

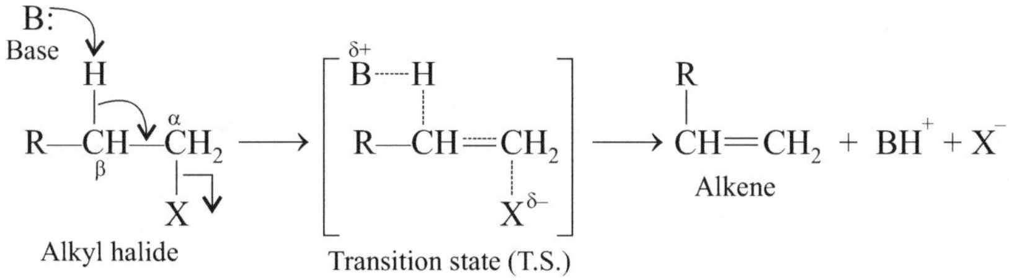

It can be noted that the reaction is a one-step process and passes through a transition state. This reaction is also known as 1, 2-elimination or simply β-elimination.

In these reactions, the two groups to be eliminated (*i.e.,* H and X) are *trans* to each other and hence E2 reactions are generally *trans*-elimination reactions.

In E2 reaction, as the number of alkyl group on the carbon with the leaving group increases, the rate of the reaction increases. Thus,

$$R_3CX \;>\; R_2CHX \;>\; RCH_2X$$
$$\text{3° Alkyl halide} \quad \text{2° Alkyl halide} \quad \text{1° Alkyl halide}$$

On the basis of the above, a 3° alkyl halide is more reactive than a 1° alkyl halide, Thus, 2-bromo-2-methyl propane (a 3° alkyl halide) is more reactive than 1-bromobutane (a 1° alkyl halide)

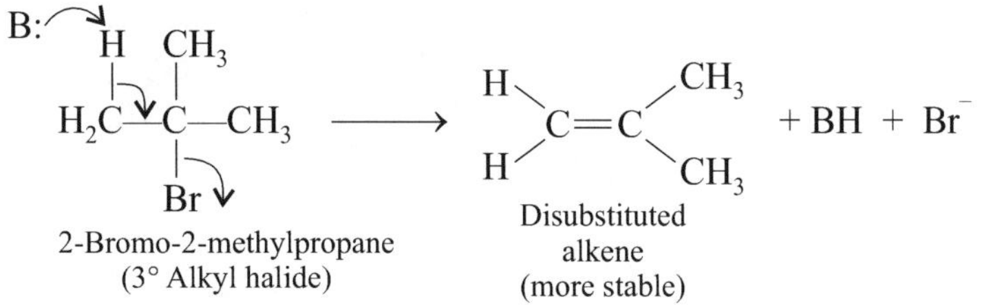

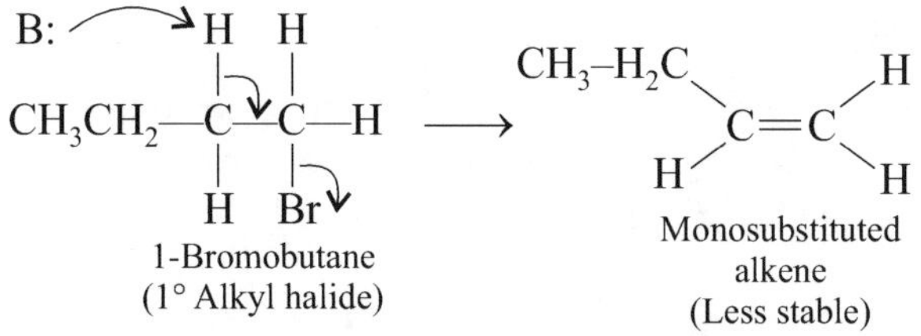

Thus, as seen a 3° alkyl halide forms a disubstituted alkene which is more stable than the monosubstituted alkene (which is less stable) formed from 1° alkyl halide.

Since the bond to the leaving group is partially broken in the transition state, the better is leaving group and faster is the E2 reaction. The order of reactivity of RX is

$$R—I > R—Br > R—Cl > R—F.$$

Also, a stronger base increase the rate E2 reaction.

An unsymmetrical substrate like 2-bromobutane on 1, 2-elimination can give either 1-butene or 2-butene

$$\overset{\overset{\displaystyle Br}{|}}{CH_3CH \cdot CH_2CH_3} \xrightarrow{base} CH_2=CHCH_2CH_3 \quad or \quad CH_3CH=CHCH_3$$

2-Bromobutane   1-Butene   2-Butene

In such cases, the formation of the olefin is determined by **Saytzeff rule**. According to this rule the major product in β-elimination gives an alkene which has more substituted double bond. Thus, in the above reaction, the major product will be 2-butene.

$$\overset{\overset{\displaystyle Br}{|}}{CH_3CHCH_2CH_3} \xrightarrow{\bar{O}C_2H_5} CH_3CH=CHCH_3 \quad + \quad CH_3CH_2CH=CH_2$$

2-Bromobutane   2-Butene 81%   1-Butene 19%
(Disabstituted alkane)   (Monosubstituted alkene)

2.6.2.1.1 Stereochemistry of E2 Reactions

In E2 reaction, the 1, 2-elimination can occur in two different ways, *viz*, syn elimination and *trans*-elimination. In *syn* elimination H and Br leave the alkyl halide from the same side, while in anti elimination both H and Br leave from the opposite side.

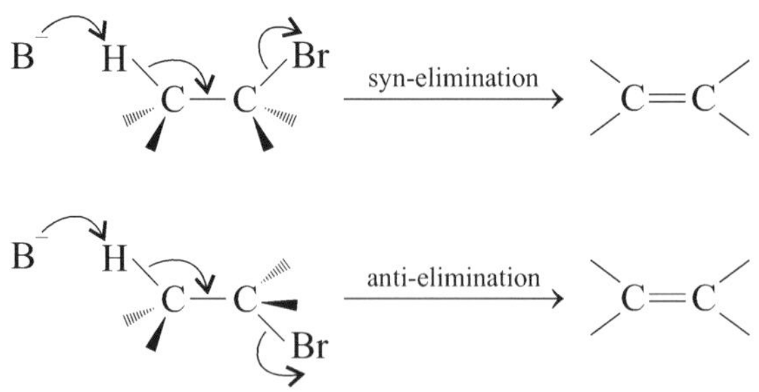

Experimentally, it is found that the E2 reaction is an anti-elimination. In an elimination reaction, the groups that are lost in the formation of the product determine the stereochemistry of the formed product.

The anti-elimination in E2 reaction is supported by the following two experiments.

(*i*) Benzene hexachloride, $C_6H_6Cl_6$ which is known to be obtained by Free radical chlorination of benzene is known to exist in eight isomeric forms, out of these forms, the one that loses HCl is 10,000 times more slowly than the others. This isomer has no adjacent chlorine and hydrogen atoms *trans* to each other.

(*ii*) The isomeric chloromaleic acid (H and Cl *cis*) and chloroformic acid (H and Cl *trans*) both give butynedioic acid (acetylene dicarboxylic acid) on treatment with a base. Out of the two substrates, chlorofurmaric acid reacts about 50 times faster.

chloromaleic acid
(H and Cl cis to each other)

chlorofumaric acid
(H and Cl trans to each other)

Butynedioic acid
(acetylene dicarboxylic acid)

So far we have seen that 1, 2-elimination occurs in alkyl halides. The process is known as dehydrobromination. 1, 2-Elimination can also occur in quaternary ammonium hydroxides. In this reaction, the base abstracts a proton from the β-carbon atom with simultaneous expulsion of a tertiary base from the α-carbon atom resulting in the formation of a double bond.

Using a quaternary ammonium hydroxide (structure given below), the base abstracts a proton from the α-carbon atom with simultaneous expulsion of a tertiary base from the α-carbon atom giving rise to the formation of a double bond.

There is one more possibility

In such a case, where there are two possibilities, the course of the reaction is predicted by **Hofmann Rule**. According to this rule, the decomposition of a quaternary ammonium salt takes place in a way so as to form a less substituted alkene. Therefore as per Hofmann rule,the reaction takes place as in the first case forming ethene.

2.6.2.2 Unimolecular Elimination Reaction (E1)

In these reactions, the rate of elimination is dependent only on the concentration of the substrate and is independent of the concentration of the nucleophile. The reaction is of the first order and is designated as E1. Like S_N1 reaction, the E1 reactions is a two-step process. The first step is the slow ionisation of alky halide to give the carbocation. The second step involves the fast abstraction of a proton by a base from the adjacent β carbon atom of carbocation, giving rise to the formation of an alkene. Thus,

At times, depending on the structure of the alkyl halide, two isomeric alkenes can be obtained. For example,

In such cases, the course of the elimination is decided by the *Saytzeff Rule*, according to which H is eliminated from the carbon atom which has lesser number of H atoms and so the major product, in the present case, is 2-butene.

As seen, the reactivity of E1 reaction is dependent on the rate of formation of the carbocation, which depends on the stability of the carbocation. The first step in E1 reaction is the formation of a carbocation. Since a carbocation may undergo rearrangement, the product formed depends on the possibility of rearrangement.

Both E1 and E2 mechanism involve the same number of bonds broken and formed. The difference between the two is that in E1 reaction, the leaving group comes off to form a carbocation before the β-proton is removed. This reaction proceeds in two steps. However, in an E2 reaction, the leaving group and the β-proton are simultaneously removed. The E1 reaction occurs via the formation of a transition state.

As in the case of E2 reaction, in E1 reaction also, the rate of the reaction increases as the number of alkyl groups on the carbon atom bearing the leaving group increases. Thus,

$$RCH_2—X < R_2CH—X < R_3C—X$$
1° Alkyl halide   2° Alkyl halide   3° Alkyl halide

Since the base does not effect the rate of E1 reaction, weak base favour E1 reaction. In fact, the strength of the base usually determines whether the reaction follows E1 or E2 mechanism. The strong base (like $\overline{O}H$, $\overline{O}R$) favour E2 reactions and the weak base (like H_2O, ROH) favour E1 reaction.

2.6.2.3 E1cB Mechanism

So fat we have seen that in alkyl halides 1, 2-elimination can take place by E2 or E1 mechanism to give alkenes. Besides alkyl halides, 1, 2-elimination can also take place in alcohols by heating in presence of acid. The process is called dehydration.

However, β-hydroxy carbonyl compounds can dehydrate in presence of a base to give α, β-unsaturated aldehydes. The mechanism by which this dehydration takes place is known as E1cB mechanism (E1cB stands for elimination, unimolecular, conjugate base). The mechanism involves loss of H and OH from α- and β- carbon to form a conjugated double bond. The dehydration of β-hydroxy carbonyl compounds take place in two steps as given below:

β-Hydroxy carbonyl compound (Aldol)   Carbanion (A)   (Resonance stabilised)

α, β-unsaturated aldehyde (croton aldehyde)

In E1cB mechanism, like in E1 mechanism, the intermediate is a carbanion.

2.6.3 Substitution Reactions

The reactions in which one or more atoms or groups of a compound are replaced or substituted by the other atoms or groups, are known as *substitution reactions*. The products obtained in such reactions are called substitution products. An example is given below:

$$CH_3—I + Cl^- \longrightarrow CH_3—Cl + I^-$$

In these reactions, one sigma bond breaks and another forms at the same carbon atom. The substitution reaction may proceed by a free radical mechanism or ionic mechanism.

2.6.3.1 Free Radical Substitution

These reactions are initiated by free radicals (see section 2.4.3.3), common examples of free radical substitution are halogenation of methane (to give carbon tetrachloride), benzene (into benzene hexachloride), toluene (into benzyl chloride) propene (into *n*-propyl bromide) and vinyl polymerisation. For details see section 2.4.3.3.

Some interesting examples of free radical substitution are allylic substitution and vinylic substitution.

Allylic Substitution

The carbon atom adjacent to a double bond is called allylic carbon atom. A typical example is that of propene, which on reaction with chlorine at high temperature (500–600°C) gives a substitution product rather than addition product.

$$CH_2 \!=\! CH—CH_3 \xrightarrow{\text{500–600°C}} CH_2 \!=\! CH—CH_2Cl$$

propene   Allyl chloride
                                (3-chloropropene)

The reaction is believed to proceeds as follows:

$$\dot{C}l : Cl \xrightarrow{\Delta} 2\dot{C}l$$

$$\dot{C}l + H\!:\!CH_2\,CH\!=\!CH_2 \longrightarrow \dot{C}H_2—CH\!=\!CH_2$$

$$\updownarrow$$

$$CH_2\!=\!CH—\dot{C}H_2$$
Allyl radical
(resonance stabilized)

$$CH_2\!=\!CH—\dot{C}H_2 + \cdot Cl \longrightarrow CH_2\!=\!CH—CH_2Cl$$
Allyl radical   Allyl chloride

Vinylic Substitution

This carbon atom attached to a double bond is called vinylic carbon. The vinylic carbon is sp^2 hybridised and the group $CH_2{=}CH{-}$ is called a vinyl group

$$CH_2{=}CH_2 + Cl_2 \xrightarrow[\text{or } hv]{500°C} CH_2{=}CHCl + HCl$$

Ethylene   Vinyl chloride

2.6.3.2 Ionic Substitution

In ionic substitution, the attacking reagents can be electrophile or nucleophile. These substitutents are of two types, *viz.*, *electrophilic substitution* and *nucleophilic substitution*.

2.6.3.2.1 Electrophilic Substitution Reactions (S_E)

These are discussed under the following heads

- Aromatic Electrophilic substitution
- Electrophilic substitution in hetero aromatic compounds

2.6.2.2.1.1 Aromatic Electrophilic Substitution

This type of substitution involves attack by an electrophile and this reaction is represented as S_E (here, S stands for substitution and E for electrophilic). The best examples are the substitution reactions of benzene, *viz.*, nitration, halogenation, sulphonation, alkylation and acylation. Let us now discuss the mechanism of these reactions one by one.

Nitration. The aromatic compounds (*e.g.*, benzene) can be nitrated with a mixture of conc. nitric acid and conc. sulphuric acid (called nitrating mixture).

$$C_6H_6 + HNO_3 \longrightarrow C_6H_6NO_2 + H_2O$$

Benzene   Nitrobenzene

The classical explanation of using sulphuric acid is that the acid absorbs the water formed in the reaction. However, this explanation is not satisfactory. It is now certain that sulphuric acid reacts with nitric acid to form a nitronium ion (NO_2^+) besides the other products.

$$HNO_3 + 2H_2SO_4 \rightleftharpoons NO_2^+ + H_3O+ + 2H\overline{S}O_4$$

In view of the above, this reaction is believed to proceed as follows:

$$\overset{..}{H}\overset{..}{O}{-}NO_2 \underset{}{\overset{H_2SO_4}{\rightleftharpoons}} H_2O^+{-}NO_2 \overset{H_2SO_4}{\rightleftharpoons} H_3O^+ + HSO_4^- + \overset{+}{N}O_2$$

Nitric acid   (protonated nitric acid
                                    + HSO$_4^-$)

H   Ḣ NO₂   NO₂

Benzene Nitronium ion   Intermediate ion   Nitrobenzene

⬡ + NO₂⁺ →(Step 1) ⬡ →(Step 2) ⬡ + H⁺

Obviously, nitration is a two-step process. Step 1 involves C—NO_2 bond formation while Step 2 involves C—H bond breaking.

Halogenation

Aromatic compounds on treatment with a halogen (chlorine or bromine) in dark and in the presence of a Lewis acid catalyst ($FeCl_3$ or $FeBr_3$) undergo halogenation. Thus, benzene on chlorination gives chlorobenzene.

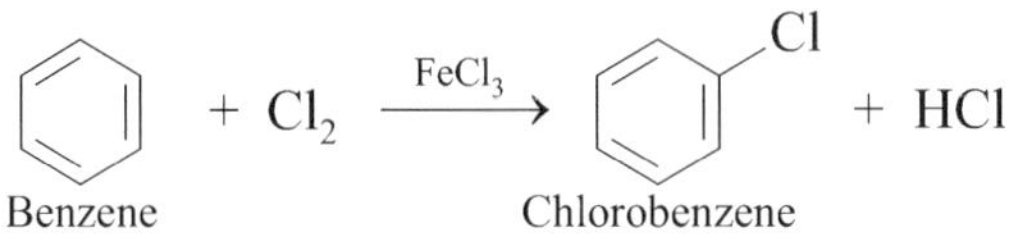

Benzene Chlorobenzene

The reaction takes place in the following three steps:

(*a*) $Cl{-}Cl + FeCl_3 \longrightarrow \overset{+}{Cl} + FeCl_4^-$

(*b*)

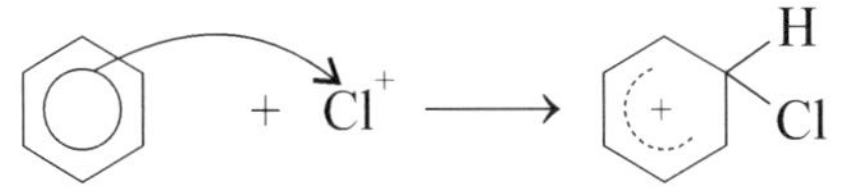

Benzene σ-Complex

(*c*)

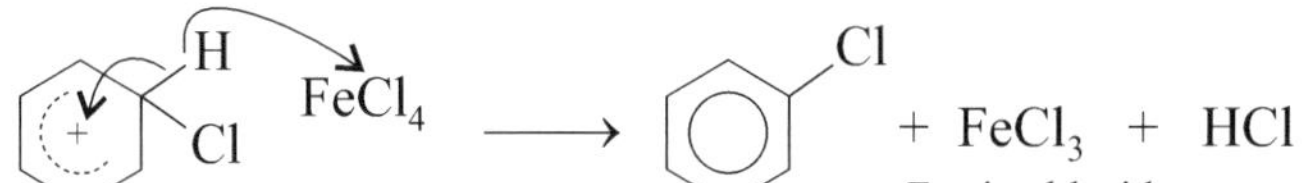

σ-Complex Chlorobenzene Ferric chloride (Regenerated)

Sulphonation

Aromatic compounds on treatment with conc. sulphuric acid or fuming sulphuric acid undergo sulphonation, giving benzene sulphonic acid.

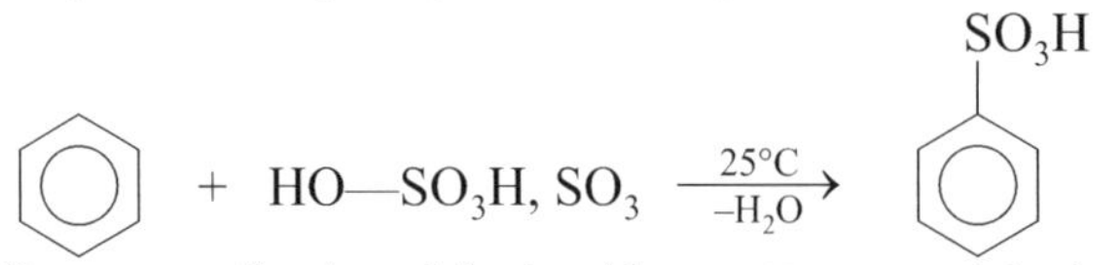

Benzene Fuming sulphuric acid Benzene sulphonic acid

Here, SO_3 acts as the electrophile and the reaction is completed as follows:

$$2H_2SO_4 \rightleftharpoons H_3O^+ + HSO_4^- + SO_3$$

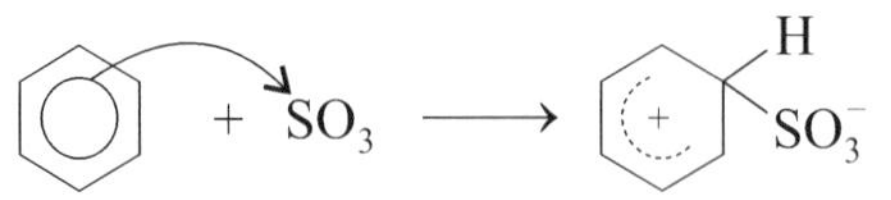

Benzene σ-Complex

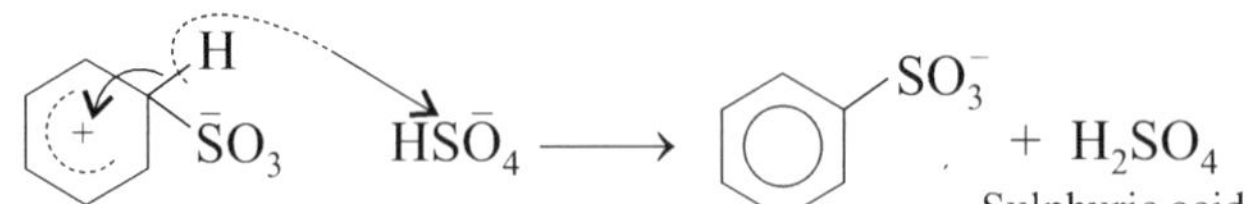

σ-Complex Sulphuric acid (regenerated)

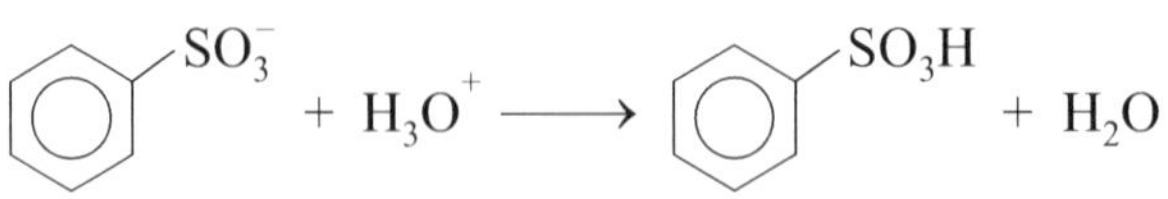

Benzene sulphonic acid

Alkylation and Acylation

Alkylation is brought about by Friedel-Crafts Alkylation and acylation by Friedel-Crafts Acylation. For details see section 2.4.3.1.

2.6.3.2.1.2 Orientation and Reactivity in Aromatic Electrophilic Substitution

We have known that in benzene, only one monosubstituted product in electrophilic substitution is obtained since all positions are equivalent. However, in case of monosubstituted benzene, the product obtained in electrophilic substitution depends on the nature of the substituent already present in the benzene ring. The presence of an activating group (*e.g.*, F, Cl, Br, I, —OH, —OCH$_3$, —OR, —NH$_2$, —NR$_2$, —NHCOCH$_3$ and alkyl groups) directs electrophilic. Substitution in ortho and para positions. On the other hand, the presence of an deactivating group (examples include NO$_2$, CF$_3$, CCl$_3$, C≡N, —SO$_3$H, —COOH etc.) directs the nucleophilic substitution in the meta position.

Thus, the alkyl group (which is known to be electron donating) pushes electron density into the ring thereby increasing the nucleophilicity of the ring the carbocation intermediate. The alkyl group being ortho and para directors direct the incoming group to ortho and para positions. This can be understood by considering the mechanism involved in generating *o*-, *m*- and *p*- isomers as shown below:

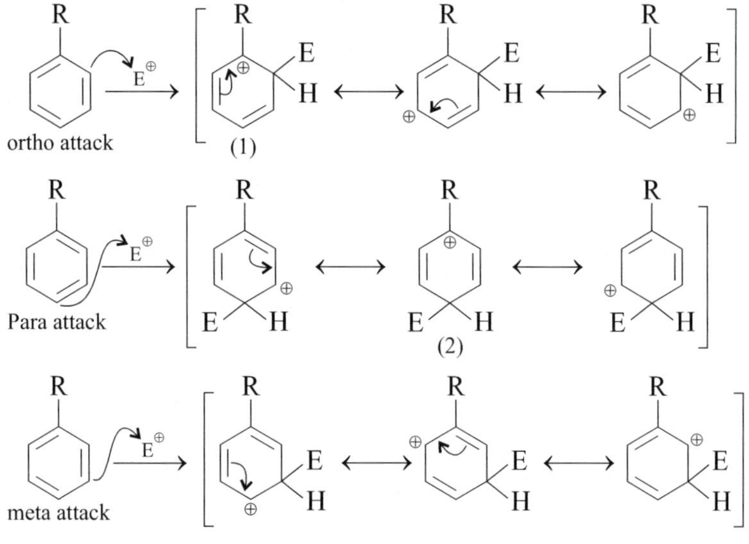

(Canonical structures arising from *o*-, *m*- and *p*- attack in case of alkyl benzene)

As seen, in case attack takes place at ortho position, out of the three canonical structures the structures (1) is most stable due to positive charge placed on carbon atom next to the alkyl group. The alkyl group in turn stabilises a neighbouring + ve charge most strongly by the inductive effect (– I).

Similarly in para attack one of the canonical structure (2) has positive charge on carbon atom adjacent to the alkyl group. There is no such structure in meta attack. So the preferred reaction pathway will be the one which goes through most stable intermediate. Thus the alkyl group is *o*- and *p*- directing.

Thus, toluene on nitration gives *o*- and *p*- nitro toluene in 59% and 39% yield respectively. A small amount (4%) of the *m*-isomers is also obtained.

Toluene $\xrightarrow[\text{H}_2\text{SO}_4]{\text{HNO}_3}$ *o*-Nitrotoluene (59%)   +   *p*-Nitrotoluene (39%)   +   *m*-Nitrotoluene (4%)

Substitutents like —OH, —OR, —NHCOR or —NR$_2$ in phenols, ethers, amides or amines activate the aromatic ring through +M (Mesomeric) effect. These groups direct the substitution to *o*- and *p*- positions. Thus, bromination of phenol give *p*- and *o*- bromophenols in the ration 2 : 1.

Phenol   + Br$_2$ $\xrightarrow[0°C]{CCl_4}$ *p*-Bromophenol (67%)   +   *o*-Bromophenol (33%)

This can be understood by comparing the intermediate canonical structures arising out of *o*-, *m*- and *p*- attack.

ortho attack   (1)   (3)

Para attack   (2)   (4)

meta attack

(Canonical structures arising from *o*-, *m*- and *p*- attack in case of phenol)

In case of phenol, the structures (1) and (2) arising from ortho and para attack have positive charge on carbon atoms adjacent to the OH group. As oxygen is nucleophilic, it can share lone pair of electrons to form a new π bond to the neighbouring electrophilic centre. This results in a fourth resonance structures (3) and (4), where positive charge is on the hetro atom. In this case, the resonance effect is more important than the inductive effect of electronegative hetero atom. Thus, further delocalization of the positive charge stabilizes the intermediate carbocation. In case of meta attack, none of the canonical structure has positive charge on carbon atom next to the substituent and the extra fourth structure is not possible.

Similarly, ortho and para orientation of —NH$_2$ group is accounted for by contributions of structures like (1) and (2). Thus, the —NH$_2$ group activates the aromatic ring and directs the substitutions to ortho and para positions.

(Canonical structures arising from *o- m-* and *p-* attack in case of aniline)

As has already been stated, the deactivating groups are meta directing. These electron withdrawing groups increase positive charge and destabilize the intermediate carbocation. This destabilization is very pronounced in the intermediates arising from *o-* and *p-* attacks and so in these cases, meta attack is favoured. This is clear by considering the canonical structures arising from *o-, m-* and *p-* attacks in case of nitrobenzene.

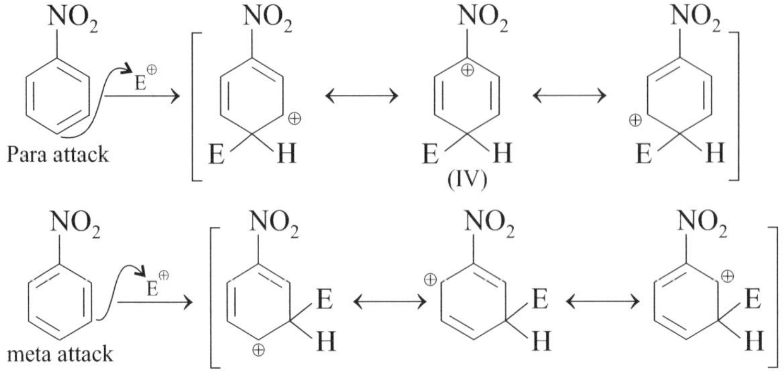

Para attack

(IV)

meta attack

(Canonical structures arising due to *o*-, *m*- and *p*- attack in case of nitrobenzene)

An interesting case is that of haloarenes. Being deactivating the halo group group is *m*- directing. However, it is *o*- and *p*- directing. In this case,the chlorine atom being highly electronegative is expected to withdraw electrons from the benzene ring and so deactivates it.

(Inductive effect of chlorine atom deactivates the benzene ring)

When attack of electrophile to chlorobenzene, takes place, the chlorine atom stabilizes the arenium ion resulting from *o*- and *p*- attack. The chlorine atom does it in the same way as an NH$_2$ group by donating an unshared pair of electrons. These electrons give rise to stable resonance structures contributing to the hybrids for *o*- and *p*- substituted arenium ions. This is clear from the following:

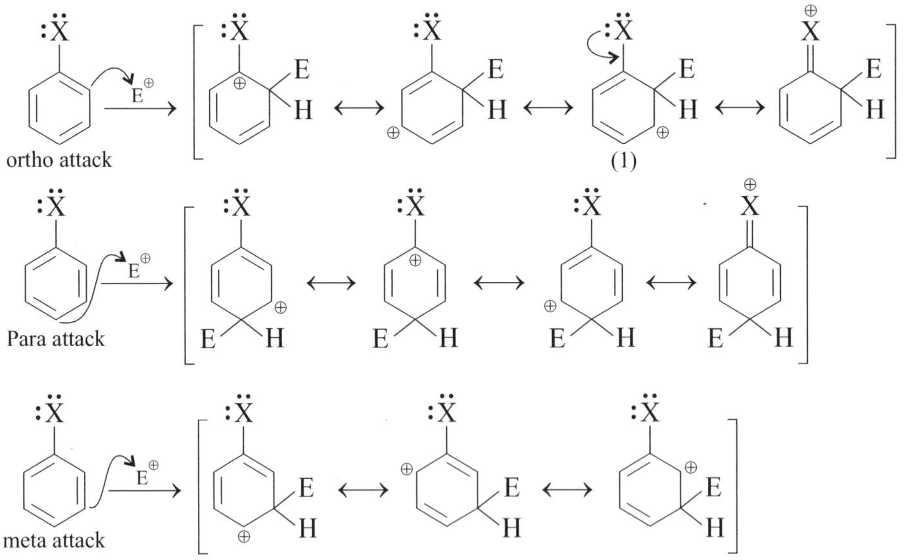

ortho attack

(1)

Para attack

meta attack

(Canonical structures arising out of *o*- *m*- and *p*- attack incase of chlorobenzene)

In case an aromatic ring has two groups out of which one group is *o*- and *p*-directing and the other group is *m*-directing, the group that is strongly activating and *o*-, *p*- directing determines the orientation of the incoming electrophile. An example is given below:

| 2-Fluoromethoxy benzene [F is m-directing and OCH₃ is *o*-, *p*-directing] | 2-Fluoro-4-nitro -methoxybenzene | 2-Fluoro-6-nitro methoxybenzene |

Steric hindrance also plays a significant role in substitution reactions. In case, a bulky substituted is present on the benzene ring, ortho attack by the electrophile is generally not possible. However if a substituent has an hetero atom, which can coordinate with the incoming group, more ortho product is produced. The increase in ortho product due to this effect at the expense of para product is called **ortho effect**.

2.6.3.2.1.3 Electrophilic Substitution in Heteroaromatic Compounds

Heteroaromatic compounds are aromatic compounds containing O, N or S as hetero atoms. Some examples of hetero aromatic compounds are given below:

| Furan | Pyrrole | Thiophene | Pyridine |

These compounds possess aromatic character and undergo electrophilic substitution reactions like nitration. Sulfonation, halogenation, Friedel-Crafts reaction etc.

The structure of the three compounds, furan, pyrrole and thiophene is considered to be as resonance hybrid of the following conaconical structures:

X = O, N or S

The order of reactivity of these three heterocyclics are: pyrrole > furan > theophene

As seen, the ring carbons in these heterocycles acquire somewhat negative charge. In all these heterocycles, the electrophilic substitution occur at position 2. This is evident by the fact that the attachment of an electrophile at position 2 afford the most stable intermediate carbocation compared to the attachment of the electrophile at position 3. This is shown in the fig below

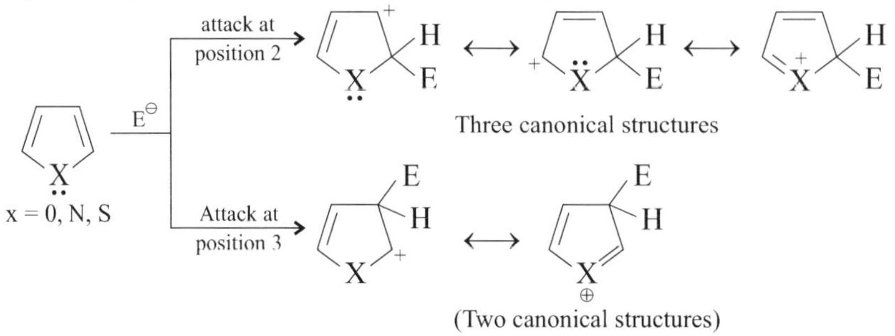

Three canonical structures

(Two canonical structures)

As seen, attack at carbon 2 produces more canonical structures compared to attack at position 3. So substitution at carbon 2 is preferred.

Following are given some of the typical electrophilic substitution reactions of furan.

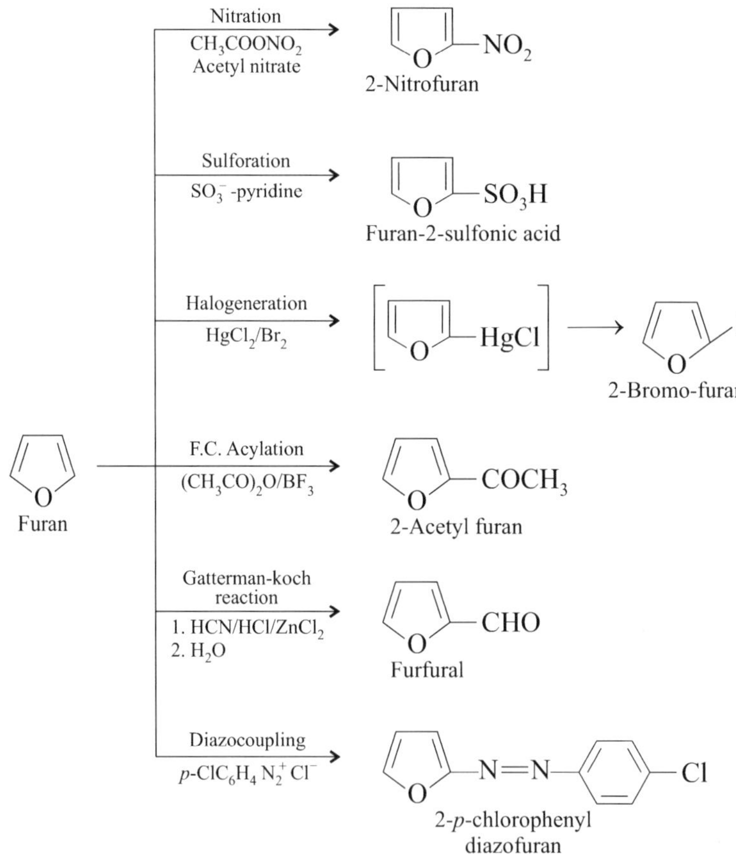

Pyrrole and thiophene also undergoes electrophilic substitution like nitration, sulfonation, Gatterman-Koch reaction, F.C. Acylation as in the case of furan.

Pyridine is a stronger base than pyrrole. This is because the sp^2 hybrid orbital of nitrogen in pyridine contains a pair of electrons which become available for bonding with acids. The structures of pyridine is best described by as a resonance hybrid of the following contributing forms.

As seen in pyridine the electronegative nitrogen produces a deficiency of electrons in the ring compared to pyrrole, where the ring carbon acquire increased electron density. It can thus be said that the nitrogen in pyridine causes deactivation of the ring. Another factor which is responsible for deactivity of pyridine ring is that under acidic conditions in most of the electrophilic substitutions (like nitration, sulphonation etc), pyridinium ion is formed which repels the positively charged electrophiles. Therefore vigorous condition are necessary for electrophilic substitutions.

In pyridine, the nitrogen atom deactivates positions 2 and 4 more than position 3. So the electrophilic substitution takes place preferably at position 3. This is well understood by consideration of the structures arising due to attack at positions 4, 3 and 2.

As seen, the attack of the electrophile at positions 4 and 2 involve a structure in each bearing a positive charge on nitrogen (structure marked A and B respectively whereas none of the intermediate structures formed from attack at position 3 bear a positive charge on nitrogen. Thus the intermediates formed by electrophilic attack at position 3 is more stable. So electrophilic substitution is favoured position 3.

Typical electrophilic substitution reactions are nitration, sulfonation and bromination, All require drastic conditions due to deactivation of the ring as already explained.

We have seen that electrophile substitution in pyridine takes preferably in position 3. However, using special technique electrophilic substitution can be made to go on position 2 or 4.

Thus, pyridine is first made to react with H_2O_2 or any other acid to give pyridine N-oxide. It is interesting to note that the presence of oxygen on nitrogen of pyridine ring reverses the direction of electrophilic substitutions of the pyridine ring. Normally, electrophilic attack on pyridine occurs in β-position. However, attack on pyridine N-oxide occurs in α- or r- positions. Thus, after the introduction of the electrophile, the pyridine oxide ring is converted into pyridine by mild reduction such as treatment with salts of iron or titanium.

2.6.3.2.2 Nucleophilic Substitutions

The substitution reaction brought about by a nucleophile is called nucleophilic substitution reaction and is denoted as S_N (here, S stands for substitution and N for nucleophilic).

Most of the work on nucleophilic substitution reactions has been carried out by Ingold and his coworkers. The best known example of such a reaction is the hydrolysis of alkyl halides with aqueous alkali to give alcohols.

$$\underset{\text{Alkyl halide}}{R\!-\!X} + \underset{\text{Aqueous alkali}}{\overset{-}{O}H} \longrightarrow \underset{\text{Alcohol}}{R\!-\!OH} + \underset{\text{Halide ion}}{X^-}$$

Following types of Nucleophilic substitutions are discussed

- Bimolecular nucleophilic substitutions (S_N2) reactions
- Unimolecular nucleophilic substitutions (S_N1) reactions
- Nucleophilic substitutions involving neighbouring group participation
- Nucleophilic substitution internal (S_Ni)

- Nucleophilic substitution on alcohols - an indirect method
- Nucleophilic substitution reactions in relatively non-polar aprotic solvent by phase transfer catalysis
- Nucleophilic Aromatic substitutions
- Nucleophilic heteroaomatic substitutions

2.6.3.2.2.1 Bimolecular Nucleophilic Substitutions (S_N2) Reactions

This reaction is of second order and its rate depends on the concentrations of both the substrate and the nucleophile. This is represented as S_N2 (S stands for substitution, N for nucleophilic and 2 for bimolecular).

Rate $\propto$ [Substrate] [Nucleophile]

Hydrolysis of primary alkyl halide (*e.g.*, methyl bromide) in aqueous base is an example of S_N2 reaction. According to Ingold, there is participation of both alkyl halide and hydroxyl ion in the rate determining step (*i.e.*, the slowest step). Ingold suggested that the reaction proceeds via a transition state in which the attacking hydroxyl ion gets partially bonded to the reacting carbon of methyl bromide before the bromide ion departs on breaking up of C—Br bond.

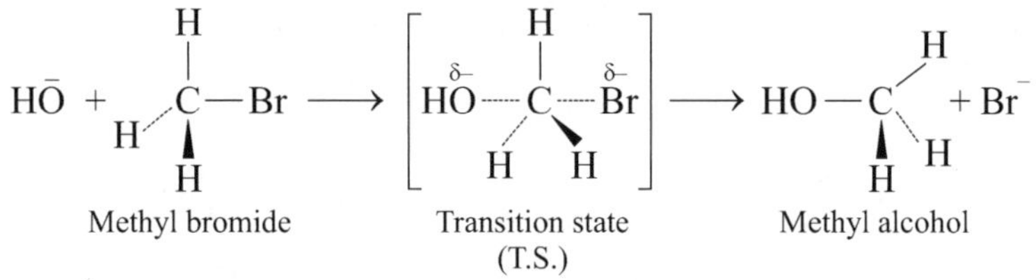

Methyl bromide   Transition state   Methyl alcohol
                                   (T.S.)

Here, the transition state is imagined as a structure in which both OH and Br are partially bonded (shown by dotted line) to the carbon atom of the methyl group. As is seen, The C—Br bond is not completely cleaved and the C—OH bond is not completely formed in the T.S. Also, the hydroxide ion has a partial negative charge (represented by δ^-) since it has begun to share its electrons with the carbon atom, and bromine too caries a partial negative charge (δ^-) because it has started moving its shared pair of electrons away from the carbon atom. According to Ingold, part of energy, necessary to affect the cleavage of the C—Br bond, comes from the energy released in forming the HO—C bond.

The stereochemical implication of an S_N2 reactions is that there is inversion of configuration.

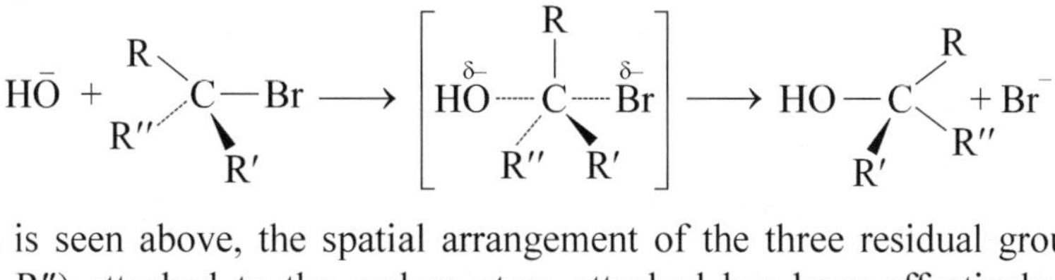

As is seen above, the spatial arrangement of the three residual groups (R, R' and R") attached to the carbon atom attacked has been effectively turned inside out, and we can say that the carbon atom has undergone *inversion* of its configuration. Thus, if we start with an optically active alkyl halide, the product (*i.e.*, alcohol) obtained will also be optically active but with opposite rotation. Thus, (+)-chlorosuccinic acid on hydrolysis with aqueous alkali gives (–) malic acid.

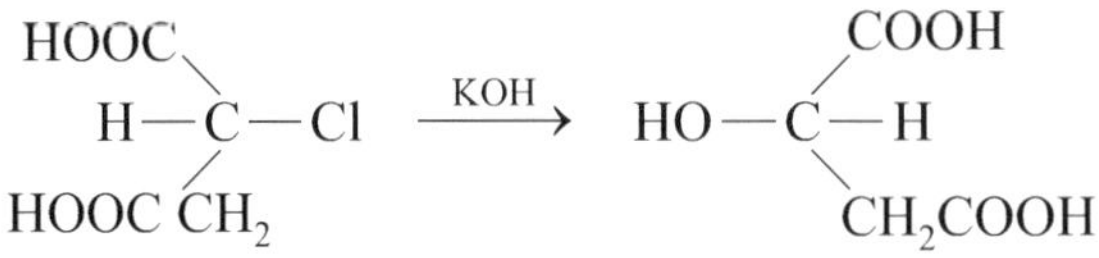

$$\underset{\text{(+)-Chlorosuccinic acid}}{\overset{\displaystyle HOOC}{\underset{\displaystyle HOOC\ CH_2}{H-C-Cl}}} \xrightarrow{\ KOH\ } \underset{\text{(–)-Malic acid}}{\overset{\displaystyle COOH}{\underset{\displaystyle CH_2COOH}{HO-C-H}}}$$

As seen in S_N2 reaction, the attack of nucleophile is from the side opposite to the leaving group. The hydrolysis of methyl bromide with alkali gives methyl alcohol as shown below:

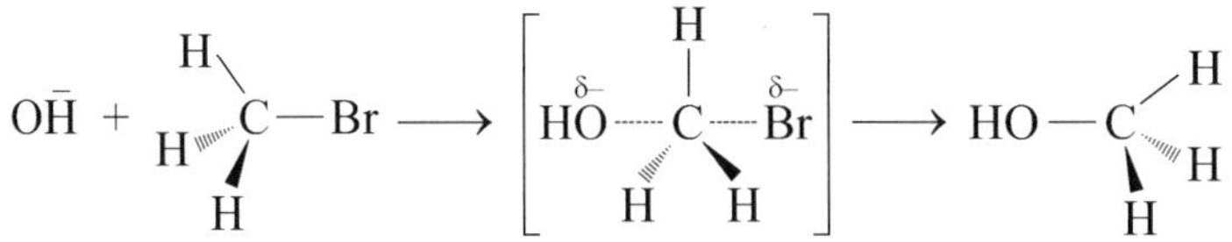

$$\overline{OH} + \underset{H}{\overset{H}{\underset{|}{C}}}\!\!-Br \longrightarrow \left[\overset{H}{\underset{H\ \ H}{HO\cdots\underset{|}{C}\cdots Br}} \right] \longrightarrow \underset{H}{\overset{H}{HO-C}}\!\!{}^{H}_{H}$$

The backside approach can also be demonstrated in the case of an optically active halide is used, an inversion of configuration can be demonstrated, As an example, optically active 2-bromooctane with NaOH gives optically active 2-octanol with inverted configuration.

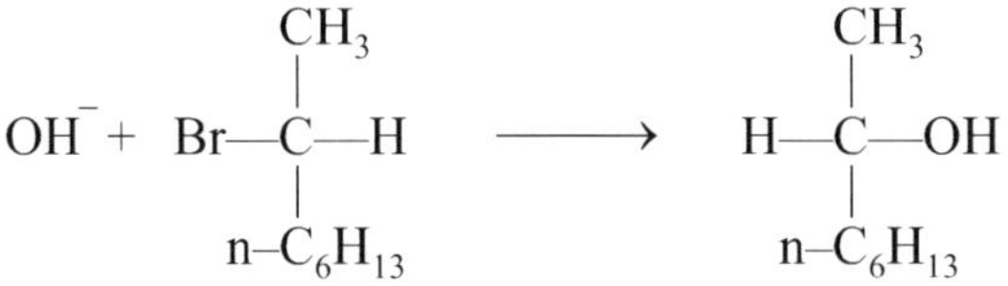

$$\overline{OH} + \underset{\underset{\text{R(–)-2-Bromooctane}}{n-C_6H_{13}}}{\overset{CH_3}{\underset{|}{Br-C-H}}} \longrightarrow \underset{\underset{\substack{\text{(S)(+)-2-Octanol}\\\text{(major product)}}}{n-C_6H_{13}}}{\overset{CH_3}{\underset{|}{H-C-OH}}}$$

The rates of an S_N2 reaction are affected by

- Nature of the substrate
- Nature of the nucleophile
- Nature of the leaving group
- Nature of the solvent

  (i) Nature of the substrate: The relative reactivity of a substrate in a S_N2 reaction depends on its structure. Table below gives the average relative rates of some alkyl substrates in a S_N2 reaction

Alkyl group	Relative rate
Methyl, CH_3^-	30
Ethyl, $CH_3CH_2^-$	1
Isopropyl $(CH_3)_2CH^-$	0.025
tert. butyl, $(CH_3)_3C^-$	~ 0
Benzyl $C_6H_5CH_2^-$	120

Among 1°, 2° and 3° halides, the reactivity is 1° halide > 2° halide > 3° halide

(*ii*) **Nature of the nucleophile:** We have sen that the rate of a S_N2 reaction depends on the concentration of both the substrate and the nucleophile. Therefore the nucleophilicity of the reagent is also an important factor in S_N2 reactions. The order of nucleophilicity of some oxygen nucleophiles is $C_2H_5O^- > OH^- > C_6H_5O^- > CH_3COO^- > H_2O$.

In case of halogens, the nucleophilicity is: $I^- > Br^- > Cl^- > F$

The nucleophilicity depends on the size of the reagent. A small nucleophile attack the carbon atom in a nucleophilic substitution with ease.

For example, methoxide anion being a small species can easily approach a carbon atom compared to *tert.* butoxide which is a poor electrophile.

(*iii*) **Nature of the leaving group:** The rate of an S_N2 reaction depends on the nature of the leaving group. Weak bases are good leaving groups and strong bases like OH^- are not. Other strong bases which are not good leaving groups are H^-, RO^-, H_2N^-, R_2N^-

Tosylate, OTS is a good leaving group and undergoes nucleophilic substitution reactions with ease. For example

$$R^+CN^- \rightarrow RCH_2 \!-\! OSO_2 \!-\!\!\langle\ \rangle\!-\! CH_3$$

$$\downarrow$$

$$RCH_2CN + R^+\bar{O}SO_2 \!-\!\!\langle\ \rangle\!-\! CH_3$$

Brosylate, OBS (*p*-bromobenzene sulphonate is also a good leaving group)

$$-OSO_2 \!-\!\!\langle\ \rangle\!-\! Br$$

(*iv*) **Nature of solvent:** S_N2 reaction takes place much faster in polar, aprotic solvents than in protic solvents. Example of protic and aprotic solvents are given below:

Prototic solvents: H_2O, HCO_2H, CH_3OH, CH_3CH_2OH, $(CH_3)_2 CHOH$, $(CH_3)_3COH$, CH_3CO_2H

Aprotic solvents: $(CH_3)_2SO$, $HCON(CH_3)_2$

In the reaction

$$CH_3(CH_2)_4CH_2X + NaCN \longrightarrow CH_3(CH_2)_4CH_2CN_7 + NaX.$$

Using cyclohexane as solvent, the yield is 71% in 20 hrs in aqueous methanol. Changing the solvent to DMSO, the yield increases to 91% and the reaction is completed in 20 min.

2.6.3.2.2.1 Unimolecular Nucleophilic Substitution (S_N1) Reactions

This reaction is of the first order (unlike the S_N2 reaction which is of the second order) and its rate depends on the concentration of only the substrate (alkyl halide) and is independent of the concentration of the nucleophile ($\overline{O}H$). It is designated as $S_N I$ (s stands for substitution, N for nucleophilic and I for unimolecular). This reaction takes place in two steps. The first step is the slow ionisation of the alkyl halide to produce the carbocation and is the rate determining step. The second step is the fast attack of the nucleophile on to the carbocation. The hydrolysis of tertiary butyl bromide to give tertiary butyl alcohol is an example of $S_N I$ reaction. The complete change is explained below:

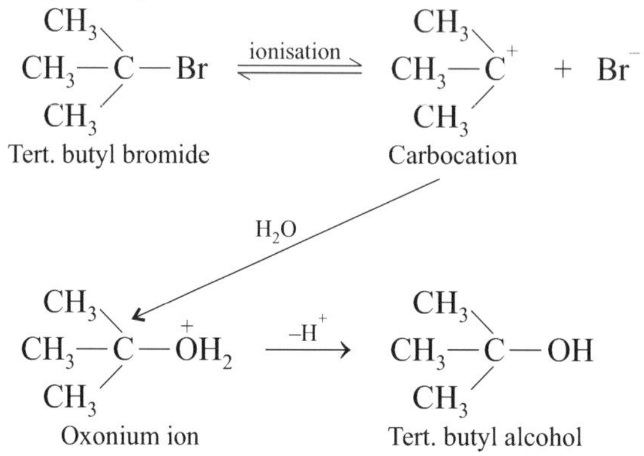

We have seen above that $S_N 2$ reaction occurs by inversion of configuration. Let us now see what happens in the case of $S_N I$ reaction. As already stated, carbocations are the intermediates (resulting by slow ionisation of alkyl halides) having a planar structure. Thus, the attack of a nucleophile on the carbocation can take place with equal ease from either side of the planar carbocation leading to a 1 : 1 mixture of the products having the same but opposite configuration of the starting material. So, racemisation will take place, yielding an optically inactive mixture of the products. This is explained as follows:

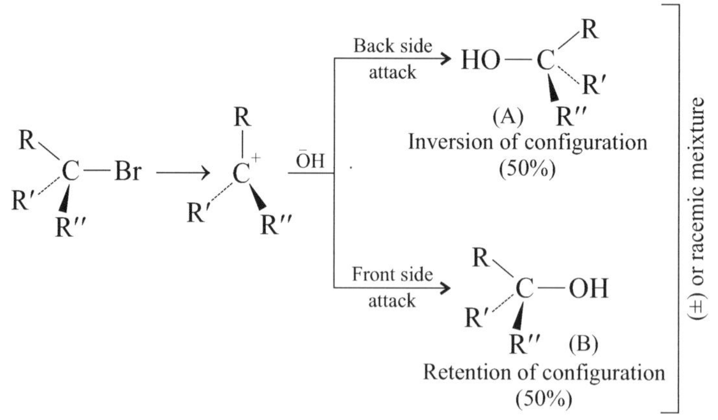

However, in practice, the product obtained is not completely racemised (±). This is because the product (A) obtained by inversion of configuration exceeds the product (B) obtained by retention of configuration. This is most likely due to the attack of the nucleophile before the departing ion has completely left the carbocation. In this way, the departing ion shields the carbocation from the frontal attack. So, the back side attack is preferred.

Factors Affection S_N1 and S_N2 Mechanisms

The course of a particular reaction *i.e.*, whether it will take place via S_N1 or S_N2 mechanism, depends on the following factors:

1. **Nature or structure of the alkyl halide:** It is found that primary alkyl halides react by S_N2 mechanism and tertiary alkyl halides react by S_N1 mechanism. However, secondary alkyl halides may react by both S_N2 and S_N1 mechanisms

$$CH_3 — CH_2 \rightarrow X$$

$$\begin{matrix} CH_3 \\ \diagdown \\ C — X \\ \diagup \\ CH_3 \end{matrix}$$

$$\begin{matrix} CH_3 \\ \diagdown \\ CH_3 \rightarrow C — X \\ \diagup \\ CH_3 \end{matrix}$$

Ethyl halide	Isopropyl halide	Tert. butyl halide
S_N2	$S_N2 + S_N1$	S_N1

As is seen, the mechanism of substitution changes from S_N2 to S_N1 as we go from left to right. This is attributed to the reason that the inductive effect of the alkyl group increases the electron density on the α carbon atom from left to right. The S_N2 reaction also slows down as the transition state (carbocation) becomes more crowded (steric factor).

2. **Nature of the nucleophile:** It is found that powerful nucleophiles (like alkoxide ions and hydroxide ions) favour S_N2 mechanism. On the other hand, weak nucleophiles (like water and alcohol) favour S_N1 mechanism. Also, higher concentration of the nucleophile favours S_N2 mechanism and low concentration favours the S_N1 mechanism.

3. **Nature of the solvent:** S_N1 mechanism is favoured as the polarity of the solvent increases.

4. **Nature of the halogen atom:** The nature of the halogen alters only the rate of the reaction which follows the order R—I > R—Br > R—Cl for both S_N1 and S_N2 mechanisms. The nature of halogen has no effect on the type of reaction mechanism.

As in the case of S_N2 reaction in case of S_N1 reaction also, the rate of the reaction depends on whether C attached to the leaving group is 1°, 2° or 3°. However, in case of S_N1 reaction, the rate of the reaction is of the order.

$$\underset{\text{Methyl}}{CH_3—X} \qquad \underset{1°}{RCH_2X} \qquad \underset{2°}{R_2CHX} \qquad \underset{3°}{R_3CX}$$

increase in rate of S_N1 reaction.

Thus, 3°alkyl halides undergo S_N1 reaction rapidly, 2° alkyl halides react more slowly and methyl and 1° halides do not undergo S_N1 reaction.

This trend is just the opposite to that observed for S_N2 reaction. The above reactivity can be explained by the ease of formation of carbocation, which is of the rate 3° > 2° > 1°.

2.6.3.2.2.3 Nucleophic Substitutions Involving Neighbouring Group Participation

In some nucleophilic substitutions it is not possible to explain the formation of the product by S_N2 or S_N1 mechanism. As an example, the substitution of bromine by methyl group involving retention in configuration cannot be explained by either a S_N2 mechanism (inversion of configuration) or a S_N1 mechanism (partial racemisation).

(L)-2-Bromopropionate   α-Lactone   (L)-2-Methoxypropionate

In such cases (as cited above), the reaction it is believed that the first step (slow), the nucleophilic carboxylate ion displaces bromine attacking the asymmetric carbon atom from backside to from a α-lactone intermediate. Subsequently in the second step the methoxide ion attack the carbon in the intermediate from the side opposite to the three membered ring resulting in retension of configuration of the substrate in the product formed. This is attributed to the participation of a neighbouring group (COO^-) in this reaction.

Another example of neighbouring group participation is provided by hydrolysis of 2-chloroethyl ethylsulphide in aqueous acetone, the reaction is about 10,000 times faster than in the case of the corresponding ether. The difference in the rate of hydrolysis cannot be explained by electronic (inductive and conjugative) or steric effect. In this case, the larger sulphur atom provides an effective neighbouring group participation as compared to oxygen in the corresponding ether.

In many reaction other groups like NH_2, NHR, NHCOR, halogen, ester (COOR), aryl (C_6H_5 and C_6H_4R) etc. act as neighbouring group.

2.6.3.2.2.4 Nucleophilic Substitution Internal (S_Ni)

In some cases of nucleophilic substitution, the substrate and the nucleophile are part of the same substrate. Such intramolecular substitutions are in general more reactive than the corresponding intermolecular reactions. Such reactions proceed by S_Ni mechanism (S_Ni implies substitution nucleophilic internal) as both the nucleophile and the leaving group are present in a single molecule. An example is given below:

$$
\begin{array}{c}
\text{CH}_3 \\
|\\
\text{H}-\text{C}-\text{O}-\text{S}=\text{O} \\
|\\
n\text{-C}_6\text{H}_{13} \quad \text{Cl}
\end{array}
\longrightarrow
\begin{array}{c}
\text{CH}_3 \\
|\\
\text{H}-\overset{\delta+}{\text{C}}\cdots\overset{\delta-}{\text{O}}\cdots\text{O}=\text{S}=\text{O} \\
|\\
n\text{-C}_6\text{H}_{13} \quad\quad \text{Cl}
\end{array}
\xrightarrow[S_N i]{\Delta}
$$

(D)-2-Octylchloro sulphite   ion pair

$$
\longrightarrow
\begin{array}{c}
\text{CH}_3 \\
|\\
\text{H}-\text{C}-\text{Cl} \quad + \quad \text{SO}_2 \\
|\\
n\text{-C}_6\text{H}_{13}
\end{array}
$$

In the above case, the chlorine from chlorosulphite attacks the asymmetric carbon from the same side leading to retention of configuration. In case, an external chloride in available in presence of pyridine, a normal S_N2 reaction occurs with inversion of configuration.

$$
\begin{array}{c}
\text{CH}_3 \\
|\\
\text{H}-\text{C}-\text{O}-\text{S}=\text{O} \\
|\\
n\text{-C}_6\text{H}_{13} \quad \text{Cl}
\end{array}
+ \text{Cl}^-
\xrightarrow[S_N2 \text{ reaction}]{\text{pyridine}}
\begin{array}{c}
\text{CH}_3 \\
|\\
\text{Cl}-\text{C}-\text{H} \quad + \text{SO}_2 + \text{Cl}^- \\
|\\
n\text{-C}_6\text{H}_{13}
\end{array}
$$

(D)-2-Octylchloro sulphite   (L)-2-Chlorooctane (inversion of configuration)

2.6.3.2.2.5 Nucleophilic Substitution on Alcohols–An Indirect Method

The OH group in alcohols is not a good leaving group. However, if the alcohols are first converted into the corresponding alkyl sulfonates, the formed alkyl sulfonate ions are excellent leaving group. Thus,

$$
\text{Nu:}^- + \text{R CH}_2-\text{O}-\overset{\overset{\text{O}}{\|}}{\underset{\underset{\text{O}}{\|}}{\text{S}}}-\text{R}' \longrightarrow \text{Nu}-\text{CH}_2\text{R} + {}^-\text{O}-\overset{\overset{\text{O}}{\|}}{\underset{\underset{\text{O}}{\|}}{\text{S}}}-\text{R}'
$$

Alkyl sulphonate (tosylate, mesylate etc.)   Very weak base –a good leaving group (sulfonate ion)

Thus, formation of a sulfonate ester is a useful way to convert an alcohol hydroxyl group in a leaving group. The sulfonate esters (*e.g.*, *p*-toluene sulphonate or methane sulfonate) are prepared by the reaction of alcohol with *p*-toluene sulfonyl chloride or methane sulfonyl chloride.

$$
\text{CH}_3 \overset{\overset{\text{O}}{\|}}{\underset{\underset{\text{O}}{\|}}{\text{S}}}-\text{Cl} + \text{H}-\text{OCH}_2\text{CH}_3 \xrightarrow[-\text{HCl}]{\text{base}} \text{CH}_3 \overset{\overset{\text{O}}{\|}}{\underset{\underset{\text{O}}{\|}}{\text{S}}}-\text{OCH}_2\text{CH}_3
$$

Methanesulfonyl chloride   Ethanol (an alcohol)   Ethyl methane sulfonate (ethyl mesylate)

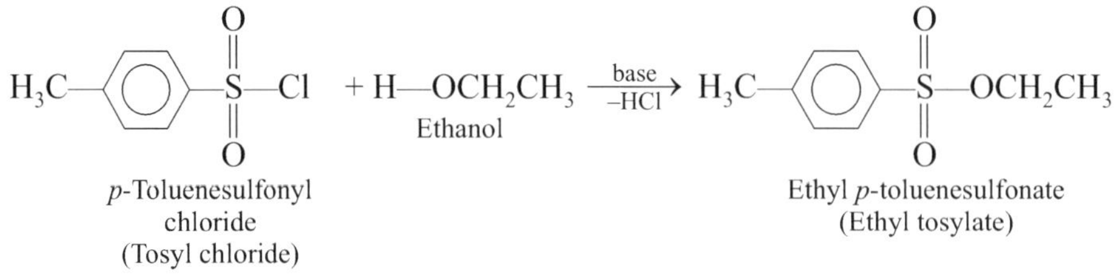

p-Toluenesulfonyl
chloride
(Tosyl chloride)

Ethyl p-toluenesulfonate
(Ethyl tosylate)

The mechanism of the formation of a sulfonate ester (e.g., alkyl methane sulfonate is given below:

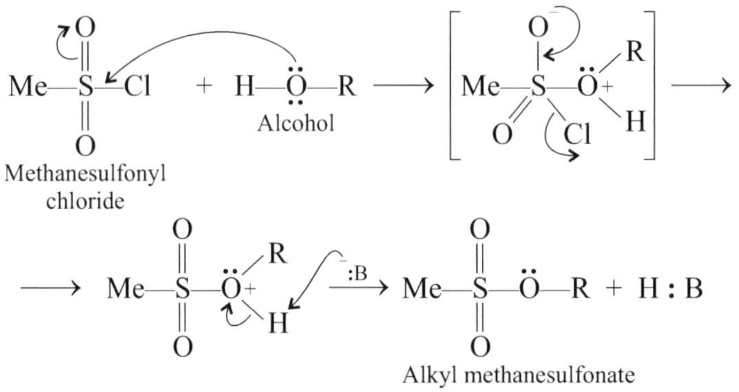

Methanesulfonyl
chloride

Alkyl methanesulfonate

The alkyl sulfonates give an indirect method for carrying out nuelophilic substitution on alcohols. In case, the carbon atom bearing the —OH group is a stereocentre, the first step viz., formation of sulfonate proceeds with retention of configuration since no bonds to the sterogoenic centre are broken. Only the O—H bond breaks. The second step proceeds with inversion of configuration.

Alkylsulfonate
(inversion of configuration)

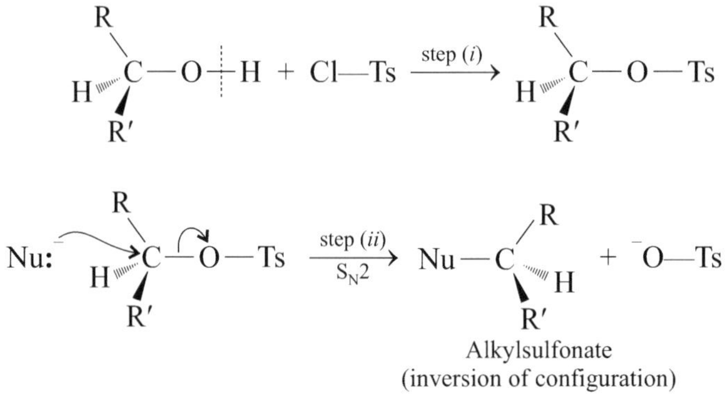

The alkyl sulfonates (tosylates etc.) undergo all the nucleophilic substitution reactions that alkyl halides do.

2.6.3.2.2.6 Nucleophilic Substitution Reactions in Relatively Non-polar Aprotic Solvents by Phase-Transfer Catalysts.

S_N2 reactions are known to take place in nucleophilic substitution reactions in polar aprotic solvents like dimethyl sulfoxide and N, N-dimethyl formamide. This is because in the polar aprotic solvents the nucleophile is only very slightly solvated and so is highly reactive. The best solvents for an S_N2 reaction are non-polar aprotic solvent such as a hydrocarbon or chlorinated hydrocarbon. However, the substrates are only slightly soluble in these solvents. This problem has been overcome by using a phase transfer catalyst, which transfers the ionic reactant into organic phase where the reaction occurs. Thus,

$$CH_3(CH_2)_6CH_2Cl \text{ in decane} \xrightarrow[aq.\,Na\,CN,\,105°C]{R_4\,N^+\,Br^-} \underset{95\%}{CH_3(CH_2)_6CH_2CN}$$

A number of nucleophilic substituents have carried out using the above method.

2.6.3.2.2.7 Nucleophilic Aromatic Substitutions

The aryl halides, like Vinyl halides are unreactive towards nucleophilic substitution under usual conditions. As an example, chlorobenzene on boiling for long time with NaOH does not form phenol. The lack of reactivity in case of chlorobenzene is basically due to prevention from backside attack as in an S_N reaction.

$$Nu^- \quad \text{⟨O⟩} \quad Cl \longrightarrow \text{No reaction.}$$

However, it is found that nucleophilic substituent reactions of aryl halides do occur when an electronic factor make the aryl carbon bonded to the halogen suspectable attack. In fact, nucleophilic substitution can occur if strong electron-withdrawing group are present in ortho and para to the halogen atoms. Thus,

As seen, *o*-nitrochlorobenzene requires the highest temperature (130°C) compared to 2, 4-dinitrochlorobenzene and 2, 4, 6-trinitrochlorobenzene which react at 100°C and 35°C respectively.

Thus, in the examples cited, the temperature necessary to bring about the reaction depends on the number of o- or p- nitro groups. A m- nitro group (as in m-nitrochlorobenzene) does not react.

The mechanism of the reaction in these cases follows an addition-elimination mechanism involving the formation of a carbanion (called Meisenheimer complex). Various steps involved in the mechanism is given below:

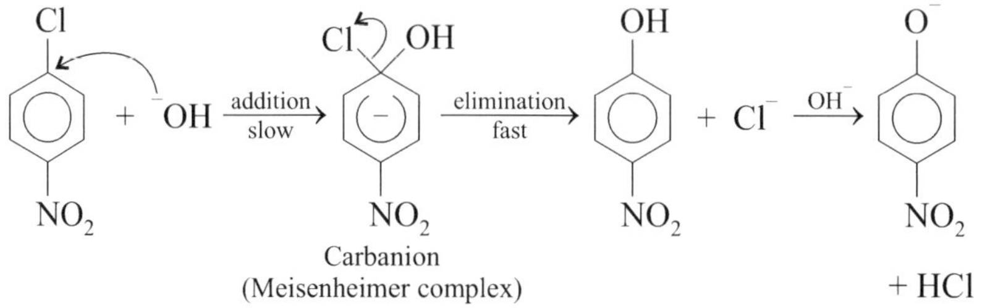

Carbanion
(Meisenheimer complex)

The above mechanism was proposed by a German chemist, Jacob Meisenheimer and is known as $S_N Ar$ mechanism.

It is believed that the carbanion formed in the above mechanism is stabilised by electron-withdrawing groups in o- and p- positions to the halogen atoms as shown below:

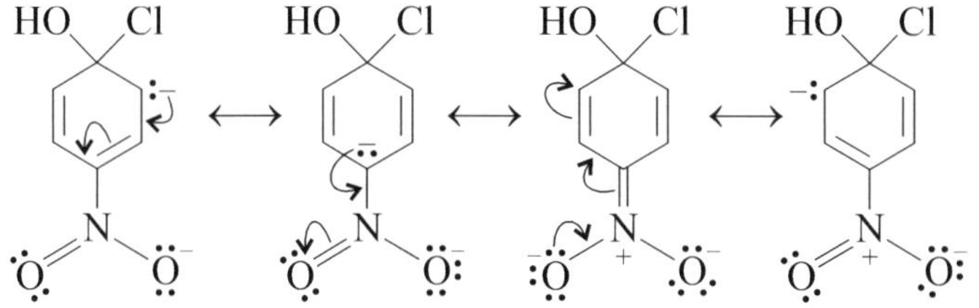

We have known that aryl halides such as chlorobenzene and bromobenzene do not react with most of the nucleophiles under ordinary conditions, they do react under drastic conditions. Thus chlorobenzene on heating with aqueous NaOH in a pressurized reactor at 350°C give phenol. This process is known as **Dow Process** and is used for industrial synthesis of phenol.

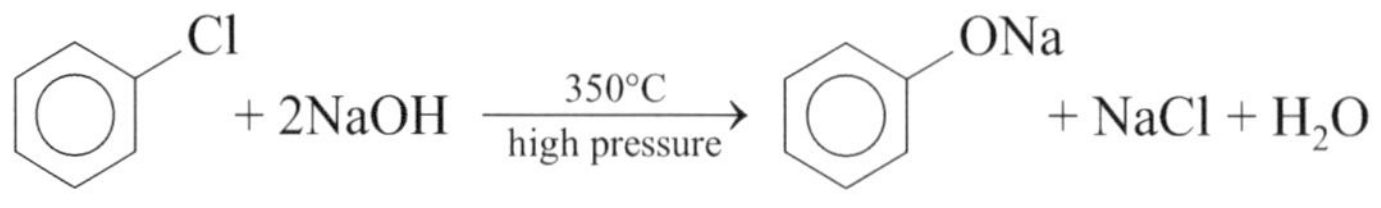

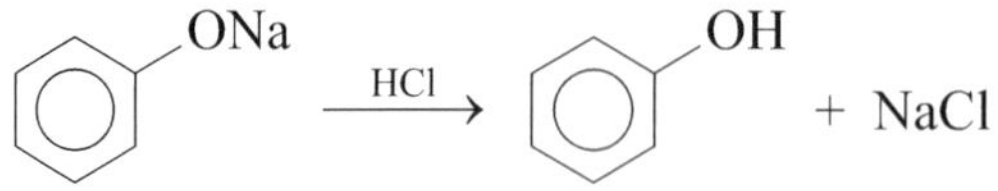

Bromobenzene reacts with only powerful base $\overline{N}H_2$ in liquid ammonia.

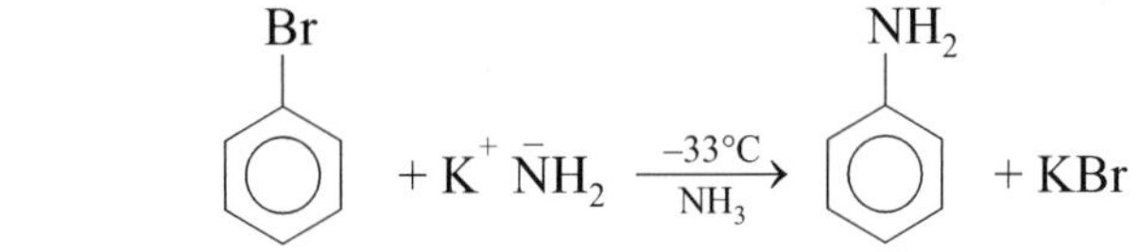

Bromobenzene $+ K^+ \overline{N}H_2 \xrightarrow[NH_3]{-33°C}$ Aniline $+ KBr$

The above reaction take place through an elimination addition mechanism involving the formation of a benzyne intermediate.

Various steps involved in the above elimination-addition mechanism are given below:

Step (*i*) Elimination

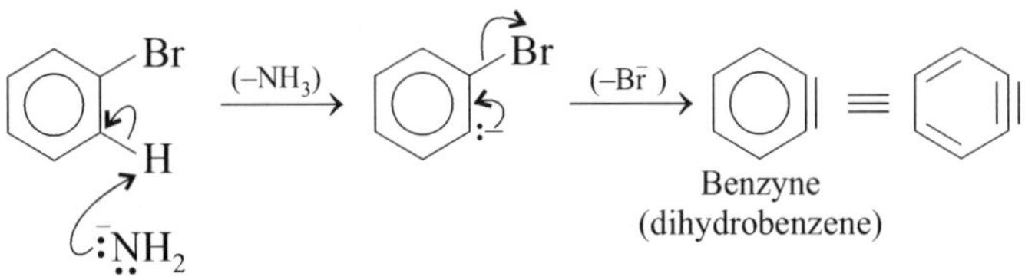

Benzyne
(dihydrobenzene)

step (*ii*) Addition

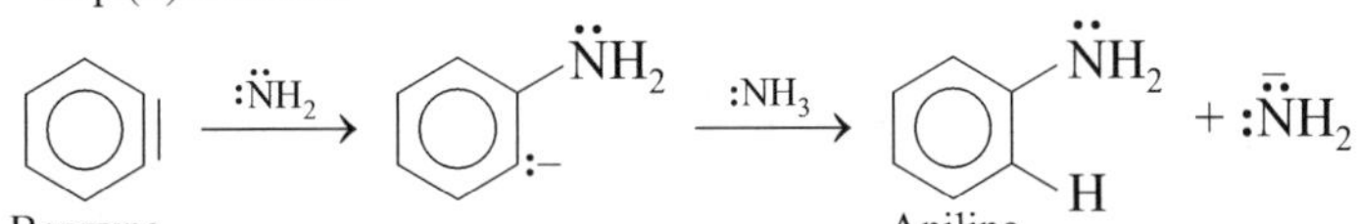

Benzyne   Aniline

The elimination-addition reaction finds support in the following experiments. It was shown (J.D. Roberts 1953) that treatment of ^{14}C labelled bromobenzene on treatment with amide ion in liquid ammonia gave aniline which had the label equally divided between 1 and 2 positions.

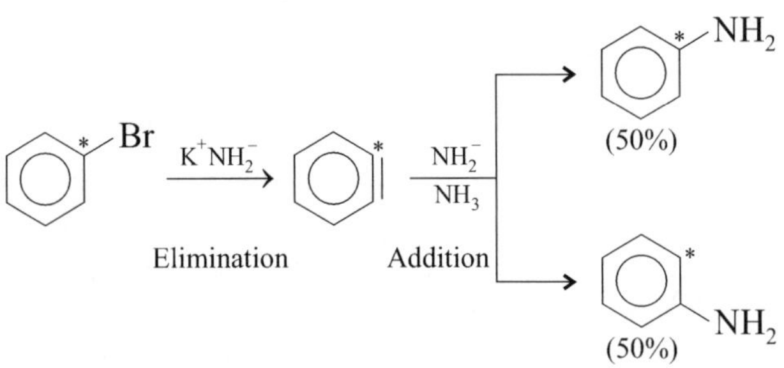

Elimination   Addition

(50%)

(50%)

Another reaction, which supports elimination-addition mechanism involves treatment of the ortho derivative (I) with $NaNH_2$ and NH_3 to give *m*-(Trifluoromethyl) aniline as the only product.

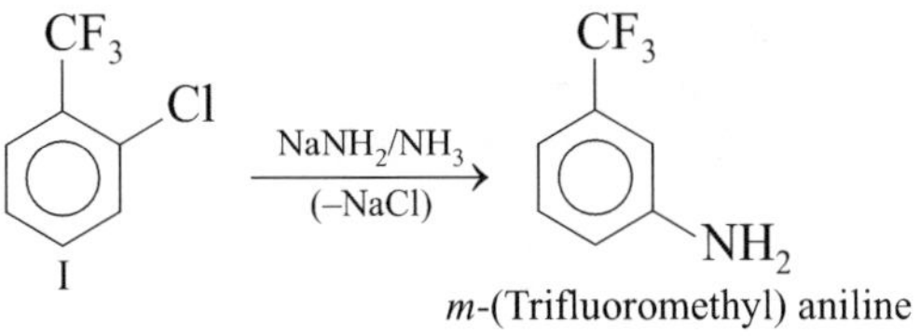

$\xrightarrow[(-NaCl)]{NaNH_2/NH_3}$

m-(Trifluoromethyl) aniline

The above result can be explained by an elimination-addition mechanism as shown below:

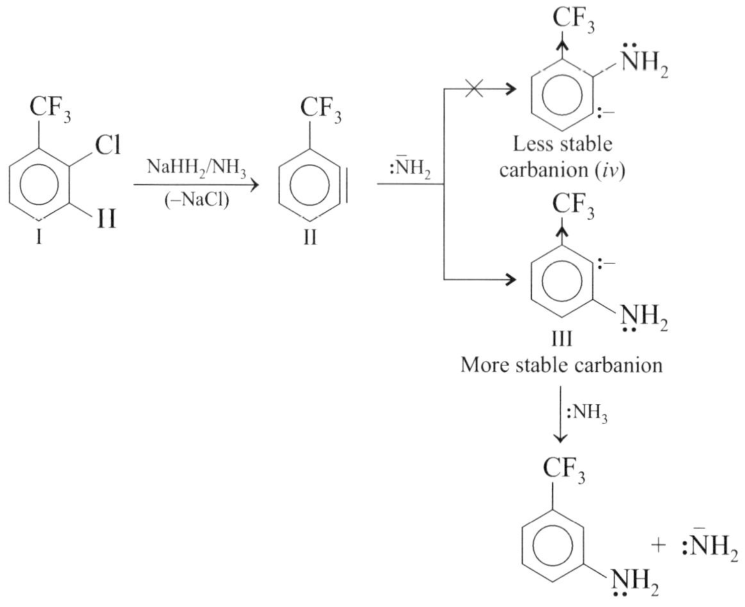

The carbanion III is more stable than the carbanion IV because the carbon atom bearing the negative charge is closer to the higher electronegative trifluoromethyl group. In fact, the trifluoromethyl group stabilizes the negative charge through its inductive effects.

2.6.3.2.2.8 Nucleophilic Heteroaromatic Substitution

We have seen that hetero aromatic compounds like furan, pyrrole, thiophene and pyridine undergo electrophilic substitutions. The present discussion relates to nucleophilic substitutions in these heteroaromatic compounds.

In furans, nucleophilic substitutions have been studied in case of halogen substituted furans. The reactivity of halofurans towards nucleophilic substitution is greater than that of the corresponding aryl halides. This is believed due to higher electron density on the carbon atom by the electron attracting tendency of oxygen. The 2-halogenfurans do not react with NaOMe at 100°C, these react with piperidine, the reaction being 10 times faster than the corresponding aryl halides.

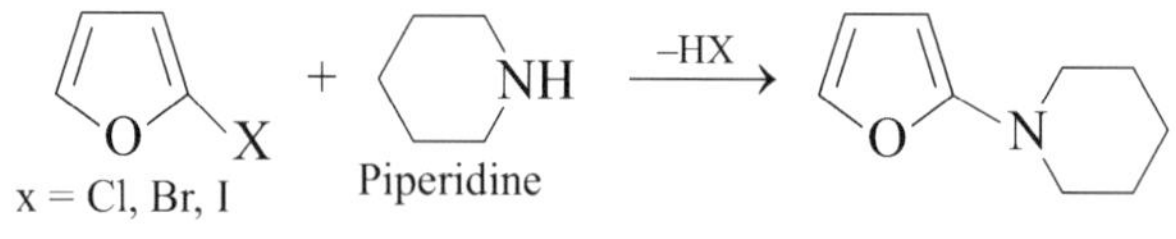

The reactivity of 2-halofuran decreases if a methyl group is present at position 3 or 5. In case of 3-iodofuran electrophilic substitution with NaOMe in presence of CuO gives 3-methoxyfuran.

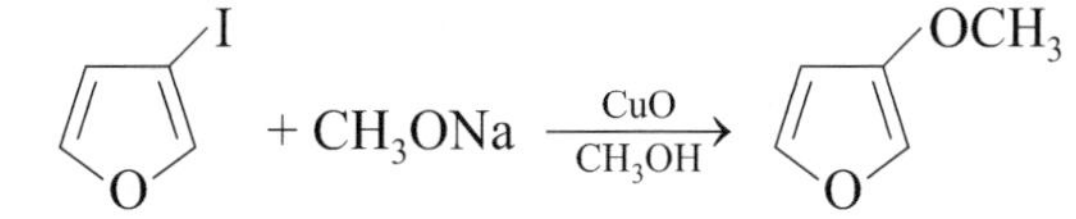

3-Iodofuran   3-methoxyfuran

The presence of electron withdrawing groups like NO_2, $COOCH_3$ or CHO facilitates nucleophilic substitution.

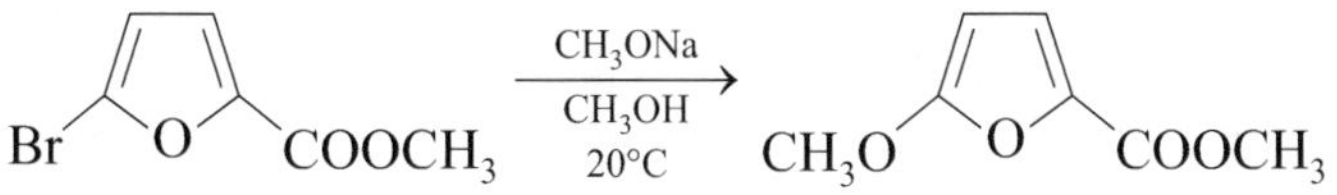

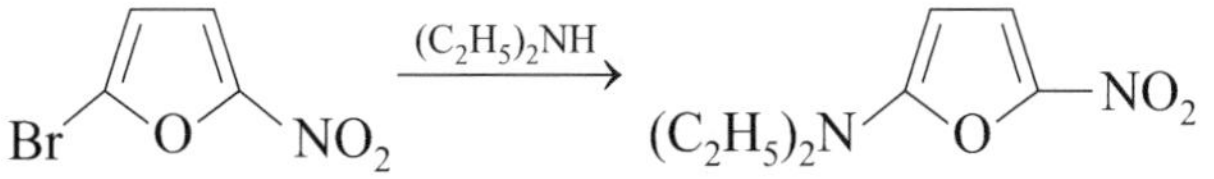

The mechanism of the above reaction involves electron transfer or radical anion as shown below:

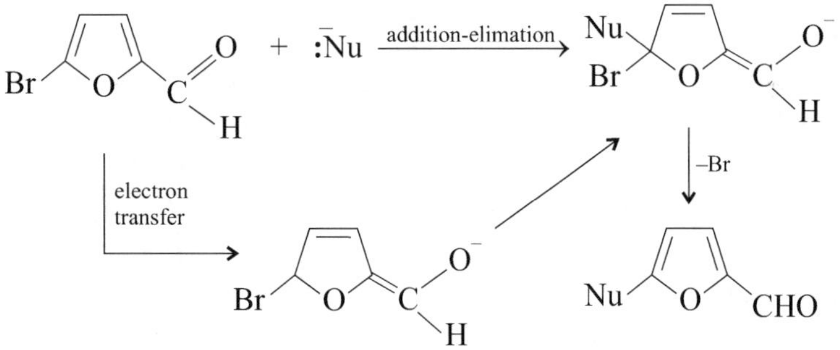

The pyrroles do not ordinarily undergo nucleophilic substitutions. This is attributed to the π-electron excessive character of pyrrole. However, nucleophilic substitutions in pyrrole take place if pyrrole is protonated or substituted by electron withdrawing groups. A typical example is given below:

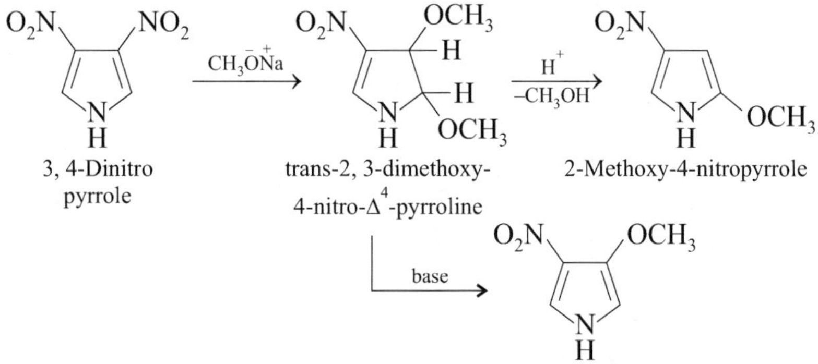

3, 4-Dinitro   trans-2, 3-dimethoxy-   2-Methoxy-4-nitropyrrole
pyrrole   4-nitro-Δ^4-pyrroline

3-Methoxy-4-nitropyrrole

The former substitutions involving formation of 2-methoxy-4-nitropyrrole is called cine-substitution and the later involving formation of 3-methyoxy-4-nitropyrrole is called ipso-substitution.

In case of thiophenes, the halothiophenes are comparatively inert towards nucleophilic substitutions. However, their reactivity is greater than aryl halides. The presence of electron-withdrawing groups in thiophene increases nucleophilic substitutions. It is found that in some cases, nucleophilic substitution does not follow the normal course of substitution involving introduction of nucleophile on the carbon atom adjacent to the carbon bearing leaving group. Thus, 2-bromothiophene on reaction with $NaNH_2$ in liquid ammonia yields 3-bromothiophene, but with KNH_2 in liquid ammonia yields 3-aminothiophene.

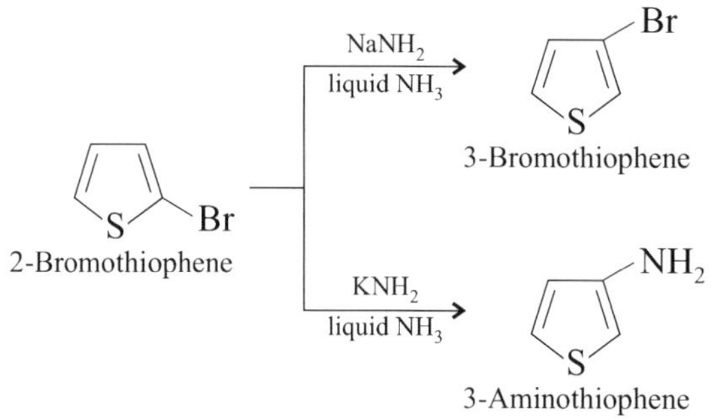

The formation of 3-bromothiophene from 2-bromothiophene by treatment with $NaNH_2$/liquid NH_3 (as shown above) involves a number of steps leading to the rearrangement and to an amination with KNH_2/NH_3 as shown below:

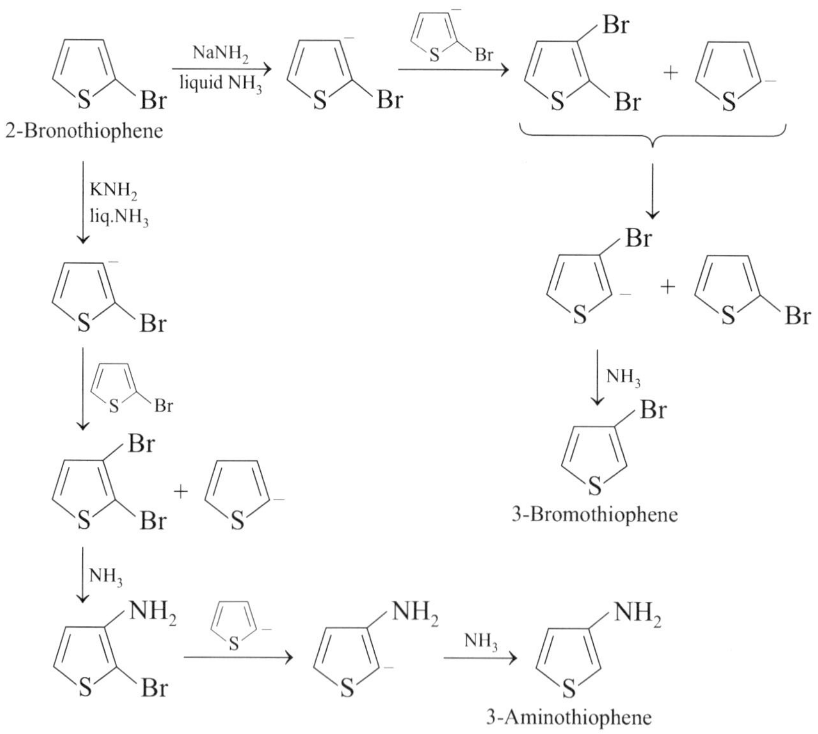

Nucleophilic substitution in unactivated halothiophenes is difficult, but can be effected under drastic conditions using copper salts as catalysts. Some examples are given below:

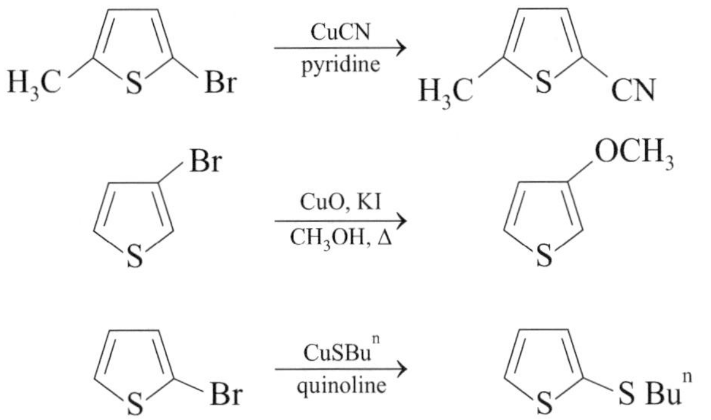

Nucleophilic substitution in pyridine can be easily affected than the five numbered heteroaromatic compounds. In case of pyridine, nucleophilic substitution takes place at 2- and 4- positions. This is because the addition of nucleophile at 2- or 4- position (but not 3-position) permits negative charge to reside on nitrogen rather than an carbon (structures I and II) as shown below:

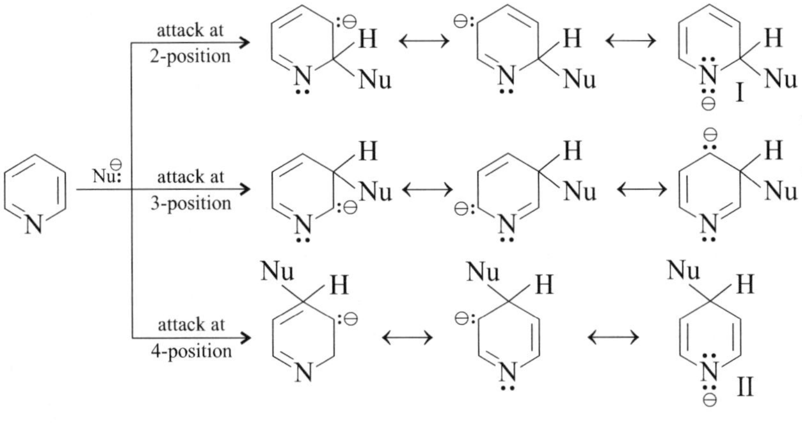

The reaction of pyridine with $NaNH_2$ (**Tschitschibabin reaction**) gives 2-amino pyridine.

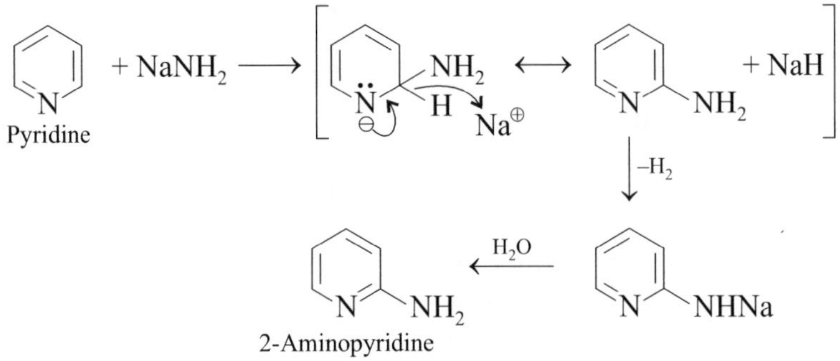

2-Aminopyridine

Some other nucleophilic substitution reactions of pyridine are given below:

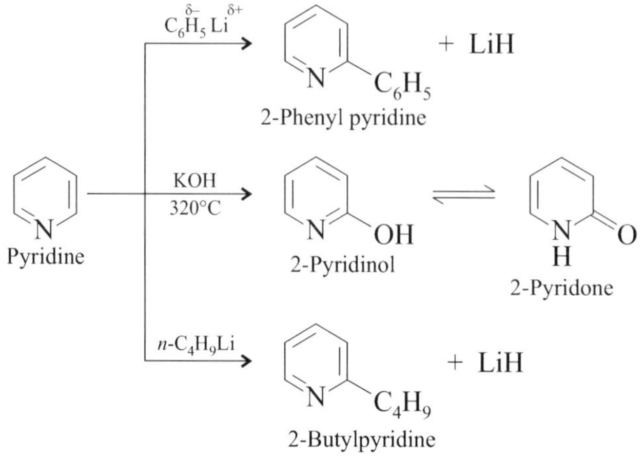

2-Phenyl pyridine

+ LiH

Pyridine

KOH
320°C

2-Pyridinol

2-Pyridone

n-C$_4$H$_9$Li

2-Butylpyridine

+ LiH

The reaction of 2-bromo or 4-chloro pyridine with ammonia at 200°C give the corresponding amino pyridine. The reaction involved nucleophilic substitution.

2-Bromopyridine   + NH$_3$ $\xrightarrow{200°C}$ 2-Aminopyridine

4-Chloropyridine   + NH$_3$ $\xrightarrow{200°C}$ 4-Aminopyridine

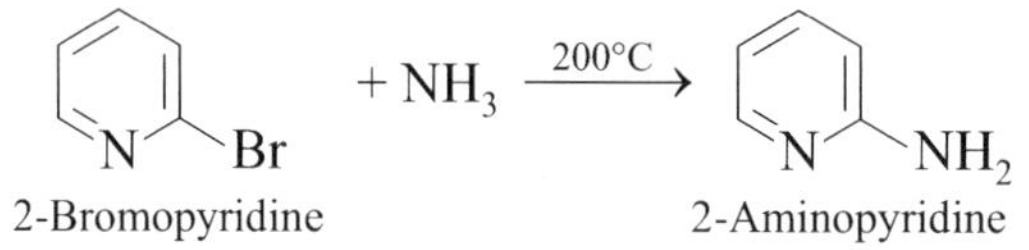

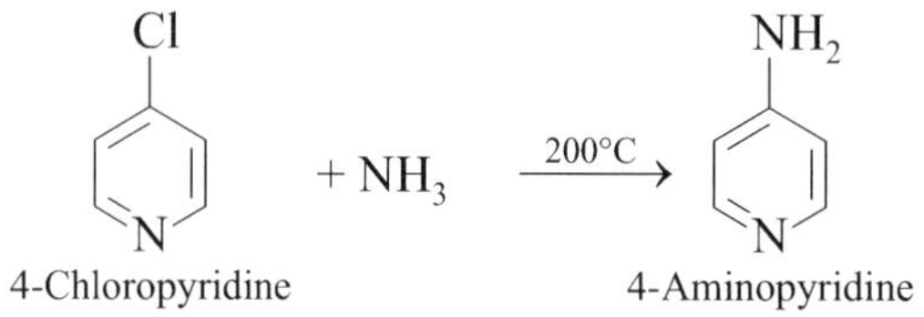

3-Halopyridines are virtually inert towards direct nucleophilic substitution. However, these are attacked by strong basic nucleophiles to give pyridyne intermediate, which an reaction with NH$_3$ give 4- and 3- aminopyridines.

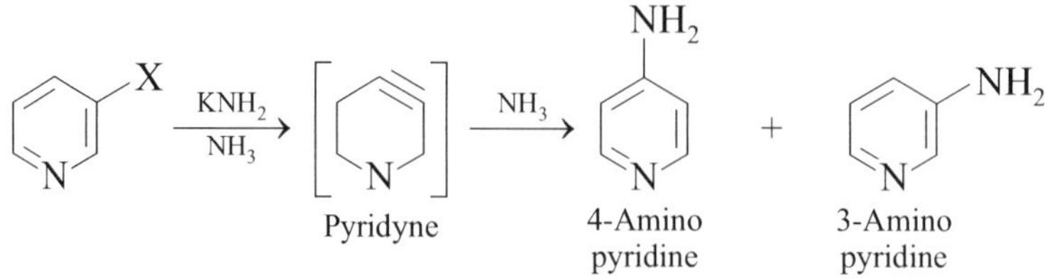

Pyridyne   4-Amino pyridine   +   3-Amino pyridine

2.6.4 Rearrangement reactions

Reactions in which some atoms or group shift from one position to another within the substrate giving a product with a new structure are called rearrangement reactions. In such reactions only the molecular skeleton is altered. These reactions may proceed by an intramolecular or intermolecular pathway. The rearrangements in which the migrating group is never fully detached from the substrate during the process of migration are called intramolecular reaction. On the other hand, in which the migration group completely detached from the substrate and and is later re-attached are called intermolecular rearrangements.

An example of intermolecular-rearrangement is the Beckmann rearrangement which involves treatment of a ketoxime with an acid catalyst (like H_2SO_4, polyphosphoric acid, $SOCl_2$, PCl_5, $C_6H_5SO_2Cl$ etc.) rearranges to a substituted amide.

In Beckmann rearrangement, the group R,which is *trans* with respect to the hydroxyl group, get migrated and the R radical or group is never detached completely from the substrate molecule.

An example of intramolecular rearrangement is **Diazoaminobenzene Rearrangement**. It involves treatment of diazoaminobenzene in weakly acidic media.

The mechanism of the reaction involves the following steps.

The rearrangement reactions may involve carbon-carbon rearrangements carbon-nitrogen rearrangements or carbon-oxygen rearrangements.

2.6.4.1 Carbon-Carbon Rearrangements

In these rearrangements, a carbon-carbon bond is broken in one part of the substrate molecule and re-formed at another place. An example is **Pinacol-Rearrangement**. In this rearrangement reaction, a methyl group migrates from one carbon to the adjacent carbon. A C=O bond appears in the product and a water molecule is lost from the reactant.

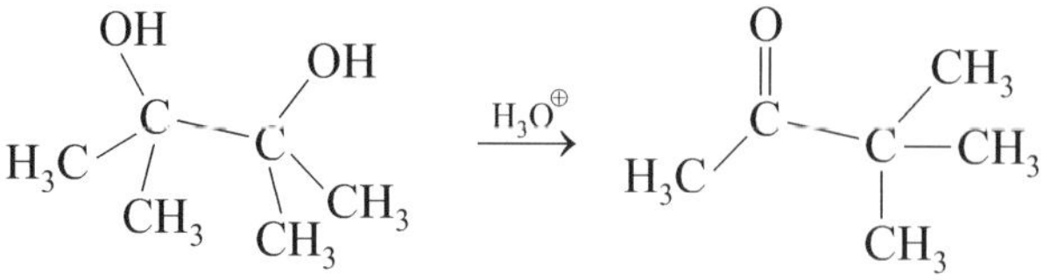

(Pinacol rearrangement)

Acid catalysed rearrangement of **pinacols** to pincolones is known as pinacol rearrangement. Thus, 2,3-dimethylbutane-2,3-diol (Pinacol) on treatment with hot 30 percent sulfuric acid gives 3,3-dimethylbutan-2-one (Pinacolone).

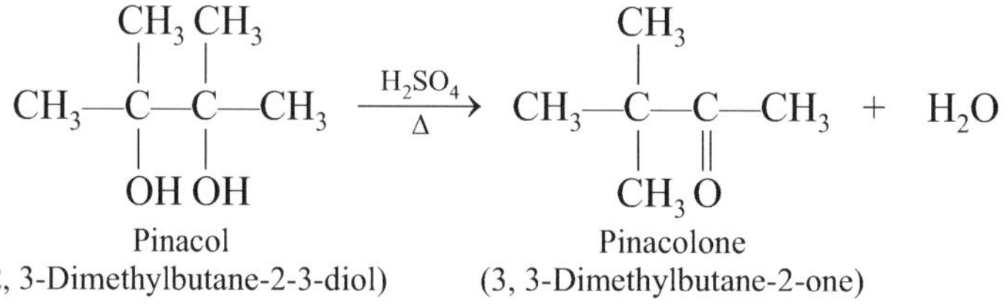

Pinacol	Pinacolone
(2, 3-Dimethylbutane-2-3-diol)	(3, 3-Dimethylbutane-2-one)

The mechanism of the rearrangement is given below

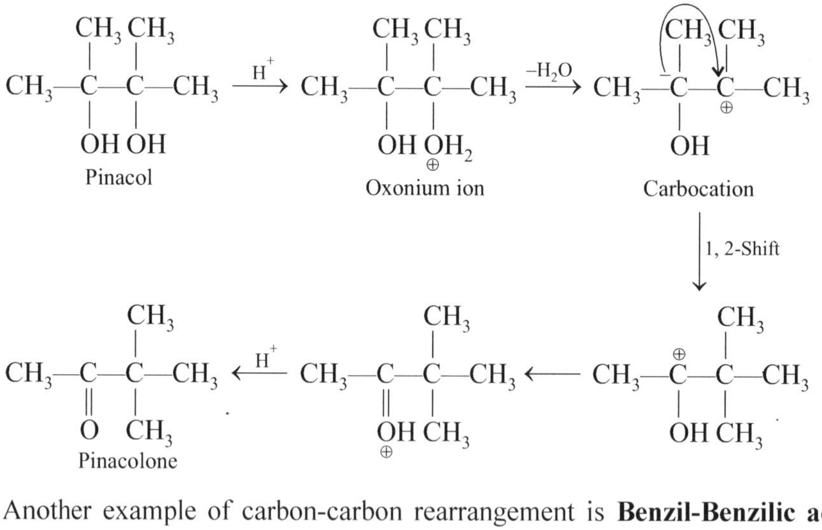

Pinacol	Oxonium ion	Carbocation

1, 2-Shift

Pinacolone

Another example of carbon-carbon rearrangement is **Benzil-Benzilic acid rearrangement**.

The α-Diketones (benzils) undergo a base catalysed reaction called benzil-benzilic acid rearrangement. Initially, the salts of α-hydroxy carboxylic acids are formed, which on acidification yield the hydroxy carboxylic acid. Thus, benzil on treatment with KOH, followed by acidification yield benzilic acid.

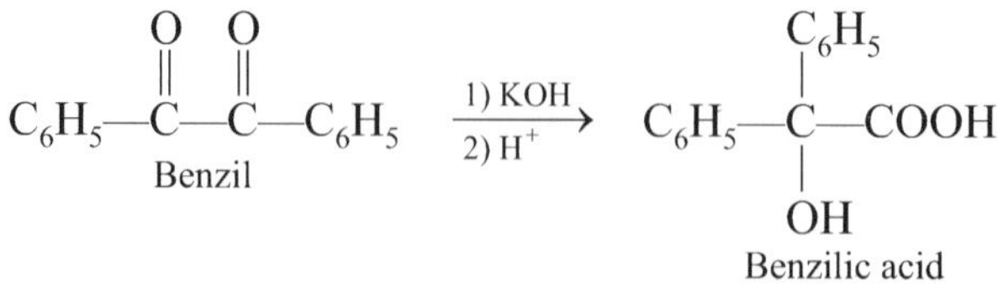

Benzil

Benzilic acid

In Benzil-Benzilic acid rearrangement there is migration of the phenyl group to the carbonyl carbon. Various steps involved in the mechanism are given below:

$$C_6H_5-\overset{\overset{O}{\|}}{C}-\overset{\overset{O}{\|}}{C}-C_6H_5 \xrightarrow[]{\overset{^-OH}{\rightleftharpoons}} C_6H_5-\overset{\overset{O}{\|}}{C}-\overset{\overset{O}{|}}{\underset{\underset{OH}{|}}{C}}-C_6H_5 \longrightarrow$$

Benzil

$$C_6H_5-\overset{\overset{\overset{-}{O}}{|}}{\underset{\underset{C_6H_5}{|}}{C}}-\overset{\overset{O}{\|}}{C}-OH \xrightarrow{\text{Proton shift}} C_6H_5-\overset{\overset{OH}{|}}{\underset{\underset{C_6H_5}{|}}{C}}-\overset{\overset{O}{\|}}{C}-O^- \xrightarrow{H^+}$$

$$\longrightarrow C_6H_5-\overset{\overset{OH}{|}}{\underset{\underset{C_6H_5}{|}}{C}}-\overset{\overset{O}{\|}}{C}-OH$$

Benzilic acid

2.6.4.2 Carbon-Nitrogen Rearrangements

It involves the migration of an alkyl or aryl group to nitrogen. Examples of such rearrangements include Beckman rearrangement (which has already been discussed in page 38) and Hofmann rearrangement.

Hofmann Rearrangement

It involves the treatment of a primary amide with bromine and hydroxide ion in water forming an amine.

$$CH_3-\overset{\overset{O}{\|}}{C}-NH_2 + Br_2 + NaOH \xrightarrow{H_2O} CH_3\,NH_2$$

Acetamide   methyl amine
                                                  1° amine

As seen, the Hofmann rearrangement results in shortening of the carbon chain by one atom and a change in functional group from an amide to an amine. Various steps involved in the rearrangement are given below.

$$CH_3-\overset{\overset{O}{\|}}{C}-NH_2 + Br_2 + NaOH \xrightarrow{H_2O} CH_3-\overset{\overset{O}{\|}}{C}-\underset{\underset{H}{|}}{N}Br + NaBr + H_2O$$

Acetamide   N-Bromamide

$$CH_3-\overset{\overset{\displaystyle O}{\|}}{C}-\overset{\overset{\displaystyle }{|}}{\underset{H}{N}}-Br \quad +\quad \bar{\;}OH \longrightarrow CH_3-\overset{\overset{\displaystyle O}{\|}}{C}-\ddot{N}\colon \; + \; Br^- + \; H_2O$$

Acylnitrine

N-Bromamide

$$CH_3-\overset{\overset{\displaystyle O}{\|}}{C}-\ddot{N} \longrightarrow CH_3\,\ddot{N}=C=O$$

Isocyanate

$$CH_3\,\ddot{N}=C=O \; + H_2O \longrightarrow \left[\, CH_3-\overset{\overset{\displaystyle }{|}}{\underset{H}{\ddot{N}}}-\overset{\overset{\displaystyle O}{\|}}{C}-OH \,\right] \longrightarrow CH_3-\ddot{N}H_2$$

Isocyanate   Carbamic acid   1° Amine
Methylamine

As seen, in Hofmann rearrangement, alkyl group migrates on to nitrogen.

2.6.4.3 Carbon-Oxygen Rearrangements

In these rearrangement on alkyl or aryl group migrates to an oxygen atom. Examples of such rearrangements are Baeyer-Villiger oxidation and Claisen rearrangement.

Baeyer-Villiger Oxidation

The oxidation of ketones to ester with hydrogen peroxide or with peracids (RCO_3H) is known as Baeyer-Villiger oxidation. The reaction can be brought about conveniently by hydrogen peroxide in weak basic solution, peroxy sulfuric acid (caro's acid) or per acid like trifluoroacetic acid, per benzoic acid, performic acid and m-chloroperbenzoic acid. With caro's acid, the rearrangement step is much faster than with peracetic acid because sulfate is a better leaving group than acetate. The most efficient reagent is trifluoroacetic acid. A typical example of Baeyer-Villiger oxidation is the reaction of acetophenone with perbenzoic acid at room temperature to give phenyl acetate in 63 percent yield.

$$\text{Acetophenone} \xrightarrow[25°C]{C_6H_5COOOH,\ CHCl_3} \text{Phenyl acetate}$$

Acetophenone (COCH₃)   Phenyl acetate (OCOCH₃)

The mechanism of Baeyer-Villiger oxidation is believed to involve following steps:

(*i*) The carbonyl reactant removes a proton from the acid (H—A) to give the protonated carbonyl reactant (1).

(*ii*) The peroxy acid attacks the protonated carbonyl reactant (1) to give the oxonium ion (2).

(*iii*) A proton is removed from the oxonium ion (2) to give the species (3).

(*iv*) The species (3) abstracts a proton from the acid (H—A) to give the species (4).

(*v*) The phenyl group migration with an electron pair takes place (from the species 4) to the adjacent oxygen, simultaneous with the departure of RCO_2H as a leaving group to give 5.

(*vi*) Final step is the removal of a proton which results in the formation of ester.

The products of Baeyer-Villiger oxidation show that phenyl group has a greater tendency to migrate than a methyl group. Had this not been the case, the product would have been $C_6H_5COOCH_3$, and not $CH_3COOC_6H_5$. The above mechanism is supported by the observation that the labelled carbonyl oxygen atom of the ketone becomes the carbonyl oxygen atom of the ester (the ester has the same ^{18}O content as the ketone).

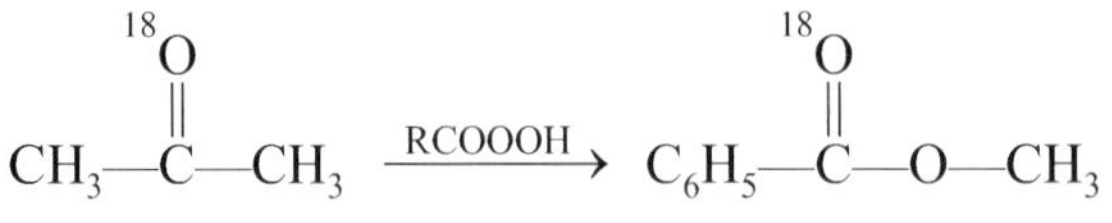

$$CH_3-\overset{\overset{\displaystyle 18}{O}}{\underset{\displaystyle \|}{C}}-CH_3 \xrightarrow{RCOOOH} C_6H_5-\overset{\overset{\displaystyle 18}{O}}{\underset{\displaystyle \|}{C}}-O-CH_3$$

Claisen Rearrangement

Allyl phenyl ether on heating to 200°C undergoes an intramolecular reaction\ called Claisen rearrangement. Claisen rearrangement is the earliest record of an organic reaction in solid state. The product is *o*-allylphenol (the allyl group migrates to the ortho position).

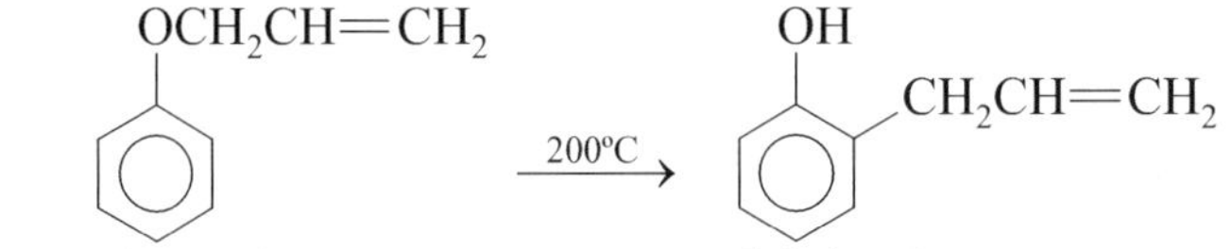

Allyl phenyl ether   *o*-Allyl phenol

The reaction does not need any catalyst and is an example of pericyclic reaction.

Mechanism

Claisen rearrangement is a [3, 3]-sigmatropic rearrangement and proceeds in a concerned manner, in which the bond between C_3 of the allyl group and the ortho position of the benzene ring form and at the same time the carbon-oxygen bond of the allyl phenyl ether breaks.

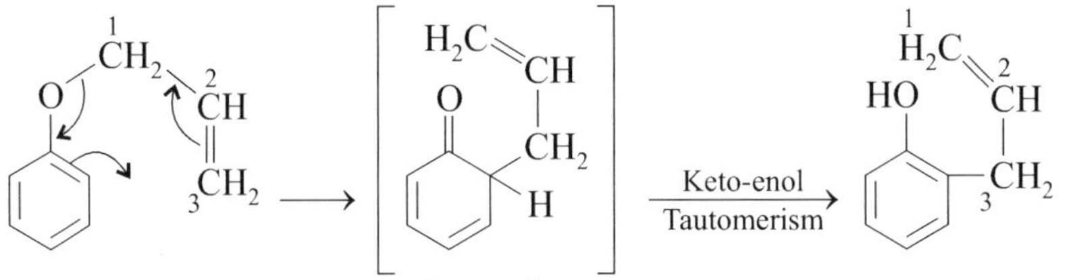

Intermediate
unstable

That only C-3 of the allyl group becomes bonded to the benzene rig has been demonstrated by carrying out the rearrangement with allyl phenyl ether containing ^{14}C at C-3. Whole of the product obtained in this reaction has the labelled carbon atom bonded to the ring.

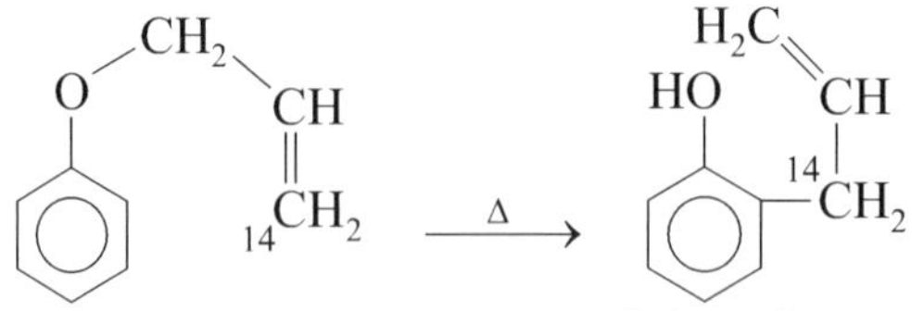

Only product

Besides the different types of rearrangement reactions (discussed above), another type of reactions, known as isomerisation reactions are also considered to be rearrangement reactions. Such reactions produce isomers of the original substrates.

2.6.4.4 Isomerisations

The isomerisation reactions are mostly intramolecular rearrangements. Some examples are given below:

(*i*) Conversion of *n*-propylbromide into isopropyl bromide. *n*-Propyl bromide on heating with ($AlBr_3$ + HBr) mixture rearranges to isopropyl bromide.

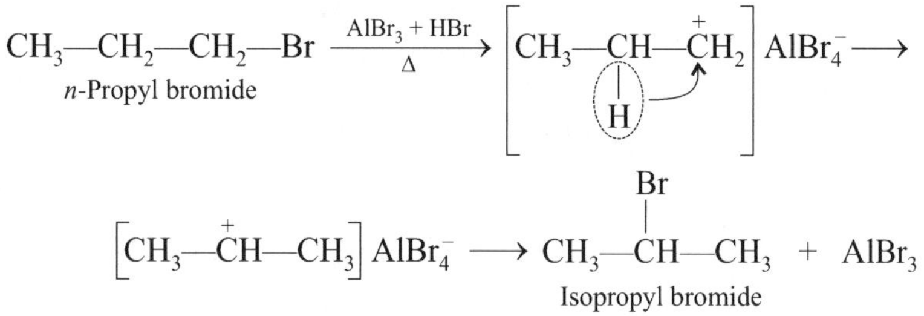

$$CH_3-CH_2-CH_2-Br \xrightarrow[\Delta]{AlBr_3 + HBr} \left[CH_3-CH-\overset{+}{CH_2} \right] AlBr_4^- \longrightarrow$$

with H below

$$\left[CH_3-\overset{+}{CH}-CH_3 \right] AlBr_4^- \longrightarrow CH_3-\overset{\overset{\displaystyle Br}{|}}{CH}-CH_3 + AlBr_3$$

Isopropyl bromide

(*ii*) **Conversion of butane to isobutane:** Butane on heating in presence of aluminium chloride gives isobutane

$$CH_3CH_2CH_2CH_3 \xrightarrow[\text{heat}]{AlCl_3} CH_3-\overset{\overset{\displaystyle CH_3}{|}}{CH}-CH_3$$

Butane   Isobutane

(*iii*) **Conversion of 1,1,1-triphenyl-2-bromoethane into 1,1,2-triphenyle-thane:** 1,1,1-Triphenyl-2-bromo ethane on treatment with lithium metal in THF followed by acidification gives 1,1,2-triphenylethane

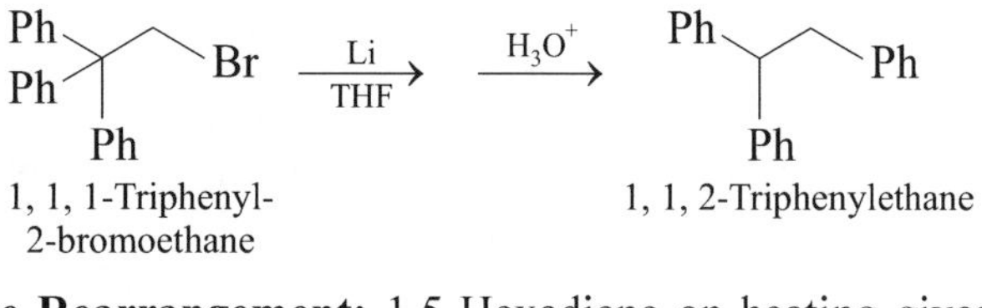

1, 1, 1-Triphenyl-
2-bromoethane

1, 1, 2-Triphenylethane

(*iv*) **Cope Rearrangement:** 1,5-Hexadiene on heating gives a degenerate 1,5-hexadiene.

1, 5-Hexadiene   Six π electron   degenerate
                         transition state   1, 5-hexadiene

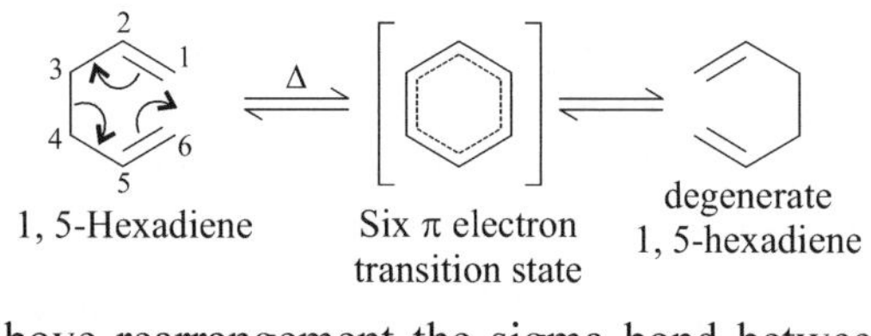

In the above rearrangement the sigma bond between C-3 and C-4 in 1,5-hexadiene reactant breaks and at the same time a new sigma bond forms between C-1 and C-6 positions in the product.

Simultaneously, both the π bonds shift and take new positions between different carbons to form a degenerate 1,5-hexadiene. This reaction is called pericyclic rearrangement.

Cope rearrangement of 3,4-dimethyl 1,5-hexadiene gives 1,6-dimethyl-1,5-hexadiene.

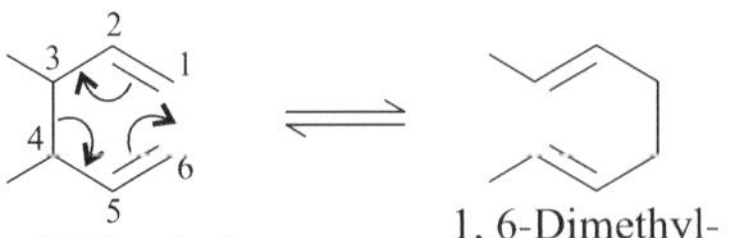

3, 4-Dimethyl-
1, 5-hexadiene

1, 6-Dimethyl-
1, 5-hexadiene

As seen in the above reaction, the reactant has monosubstituted double bonds and the product formed has disubstituted double bonds.

Another example of cope rearrangement involves thermal rearrangement of 1,5-hexadiene having a hydroxyl substituent at position 3.

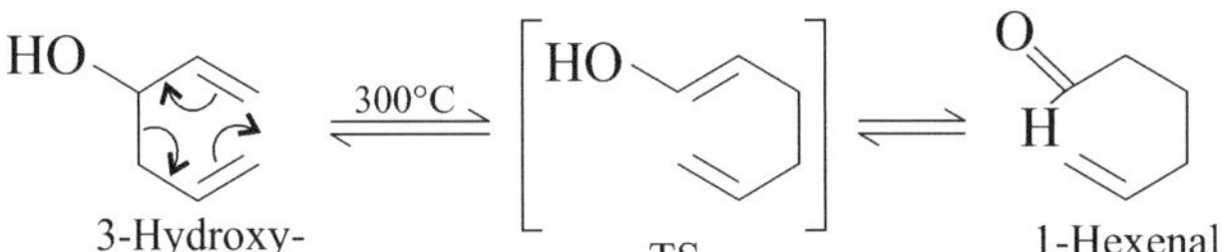

3-Hydroxy-
1, 5-hexadiene

TS

1-Hexenal

(v) **Electrolytic Reactions:** In these reactions, a single bond is formed between the termini of a conjugated polyene system. Thus, in an electrolytic reaction, intramolecular interaction of both ends of a π system leads to a intramolecular cyclisation. Thus, 1,3,5-hexatriene gives 1,3-cyclohexadiene.

1, 3, 5-Hexatriene   1, 3-Cyclohexadiene

The above electrolytic reaction like cope rearrangement also proceeds via a six-electron transition state.

2.6.5 Photochemical reactions

Photochemical reactions are known to take place almost from the time of formation of earth. As we know photosynthesis is primarily responsible in sustenance of life. In this process light energy is stored as chemical energy in plants. The photosynthesis in a simple way is represented as:

$$H_2O + CO_2 \xrightarrow[\text{chlorophyll}]{h\nu} \text{carbohydrates} + O_2$$

Another photochemical process, which in known to be taking place is the combination of ammonia, carbon dioxide and water in presence of sunlight to produce amino acids, which in turn are the building blocks of both animals and plants.

$$NH_3 + CO_2 + H_2O \xrightarrow{h\nu} \text{amino acids} + \text{protein}$$

The photochemical reactions are believed to involve absorption of electromagnetic radiation to produce electronically excited states, which in turn give the products. These reactions are in a way different from thermal reactions in that the products obtained are different in a number of cases than those obtained by thermal reactions.

Following are given some of the interesting photochemical reactions.

2.6.5.1 Free Radical Chlorination

Most of the halogenations proceed in presence of sunlight via formation of free radicals. Thus methane can be converted into carbon tetrachloride, benzene into benzene hexachloride and toluene into benzyl chloride.

2.6.5.2 Photoreductive Dimerisation

It is well known that benzophenone on treatment with zinc and acetic acid gives benzopinacol. This reaction also occurs when a solution of benzophenone in isoproyl alcohol is exposed to sunlight. It is known as photoreductive dimerisaion.

$$(C_6H_5)_2C=O + (CH_3)_2CHOH \xrightarrow{hv} C_6H_5-\underset{\underset{OH}{|}}{\overset{\overset{C_6H_5}{|}}{C}}-\underset{\underset{OH}{|}}{\overset{\overset{C_6H_5}{|}}{C}}-C_6H_5 + CH_3COCH_3$$

Benzophenone   isopropyl alcohol   Benzopinacol

The mechanism involved in the above dimerisation is given below:

$$(C_6H_5)_2CO \xrightarrow[n-\pi^*]{hv} (C_6H_5)_2 = 0 \xrightarrow[\text{crossing}]{\text{intersystem}} (C_6H_5)C=0$$
Benzophenone S_1 T_1

$$Ph_2C=0 + (CH_3)_2 \overset{H}{\underset{}{C}}=OH \longrightarrow (C_6H_5)_2 \overset{\cdot}{C}OH + (CH_3)_2 \overset{\cdot}{C}OH$$
T_1 isopropanol   benzhydrol radical

$$(C_6H_5)_2C=0 + (CH_3)_2 \overset{\cdot}{C}OH \longrightarrow (C_6H_5)_2 \overset{\cdot}{C}OH + CH_3COCH_3$$
S_0

$$2(C_6H_5)_2\overset{\cdot}{C}OH \longrightarrow C_6H_5-\underset{\underset{OH}{|}}{\overset{\overset{C_6H_5}{|}}{C}}-\underset{\underset{OH}{|}}{\overset{\overset{C_6H_5}{|}}{C}}-C_6H_5$$
Benzopinacol

2.6.5.3 Photomerisation of *cis*- and trans-1,2-diphenylethene (Stilbene)

Olefins are known to exhibit geometrical isomerism. The photoisomerisation of *cis-trans* isomerisation of stilbene (1,2-dipheylethene) represents the simplest case of light induced geometrical isomerisation.

Thus, irridation of a solution of trans stilbene in hexane with UV light results in the formation of cis-isomer. After some time of irridation the cis-trans ratio becomes constant. This condition is called a photo stationary state. It has been observed that the equilibrium favours the formation of the *cis*-isomers.

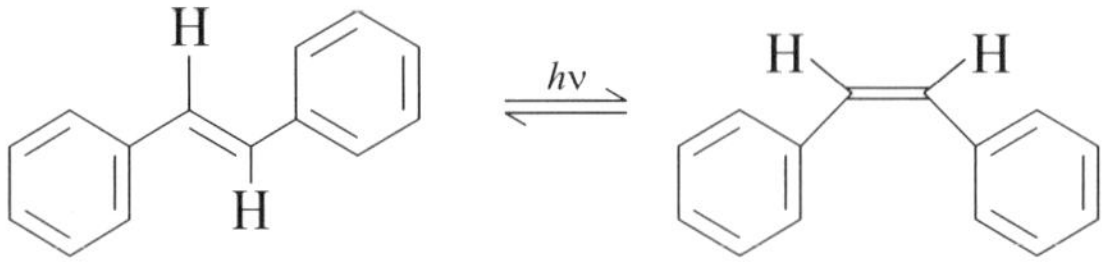

trans-stilbene
$(\lambda_{max}(\varepsilon)) = 295$ nm (16300)

cis-slibene
$(\lambda_{max}(\varepsilon)) = 276$ nm (2280)

It is formed that the trans isomer is more stable. However, thermal equilibrium favour the conversion of *cis*-stilbene to *trans*-isomer.

2.6.5.4 Photochemical Cyclo Addition Reactions

The [2 + 2] and [4 + 2] cycloadditions occur photochemically with or without sensitizers. Examples of both types of cycloadditions are given below:

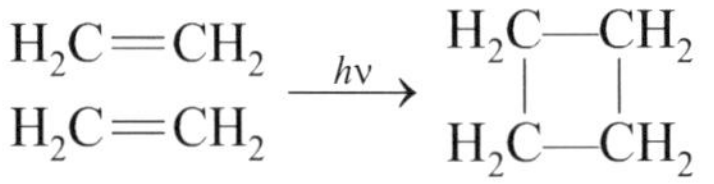

(2 + 2-cycloaddition)

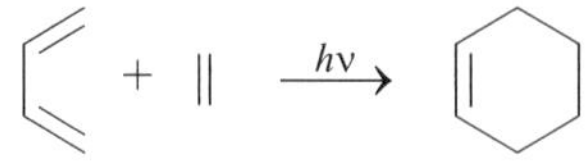

(4 + 2-cycloaddition)

Photochemical cycloaddition of olefins give four-membered rings in a synthetically useful process. A typical example in the dimersiation of cyclopentenone on irridation with light in dichloromethane give a mixture of 'head to head' and 'head to tail' isomers. These dimers are believed to be formed via an examer (excited dimer) derived from the $(\pi - \pi^*)$ cyclopentenone and a molecule of ground state cyclopentenone.

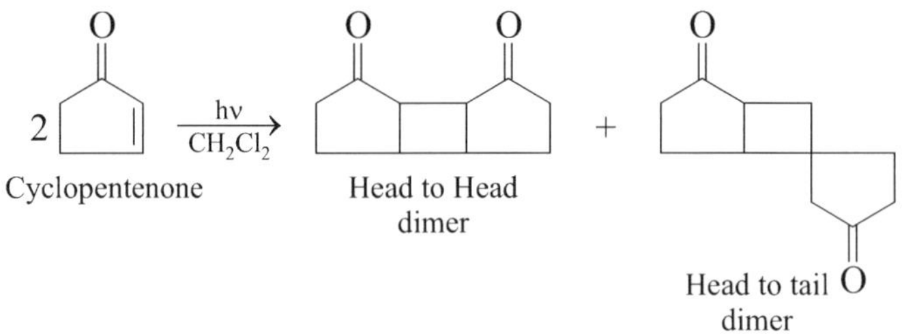

Cyclopentenone

Head to Head
dimer

Head to tail O
dimer

The photocyclisation may also proceed in an intramolecular fashion as shown below:

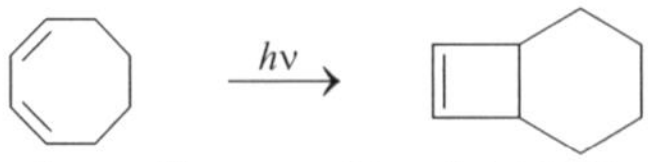

1, 3-cyclooctadiene   Bicyclo [4.2.0] oct-7-ene

Butadiene provides an excellent example to show the difference between a thermal and photochemical reaction.

It is found that the photochemical cyclo addition products of butadiene depend on the sensitizer used.

2.6.5.5 Photo chemical Erndt-Eistert Synthesis

It involves the reaction of α-diazoketone under photochemical conditions to give carbene intermediate which yields ketene. The reaction of ketene with methanol gives carboxylic ester. The overall reaction is known as Arndt-Eistert synthesis and the conversion of α-diazoketone into ketene is known as **Wolf rearrangement**.

2.6.5.6 Barton Reaction

In Barton reaction, a methyl group in the δ-position to an OH group is converted into oxime group.

Barton reaction provides an excellent route for the conversion of 18-methyl group in steroids to the corresponding oxime derivative.

In Barton reaction, the OH group on reaction with nitrosyl chloride (NOCl) gives the nitrite, which on UV irridation undergoes homolytic cleavage to give alkoxy radical. Subsequent abstraction of H atom from a carbon in the δ-position to the original OH group gives nitroso alcohol which tautomerises to oxime. Also, the oxime group can be hydrolysed to aldehyde. Thus, in Barton reaction methyl group in δ-position to an OH group is converted to aldehyde.

2.6.5.7 Photonitrosation

Cylcohexane on reaction with nitrosyl chloride in presence of UV light gives cyclohexane oxime, which is a starting material for the synthesis of caprolactam, the monomer of nylon 6.

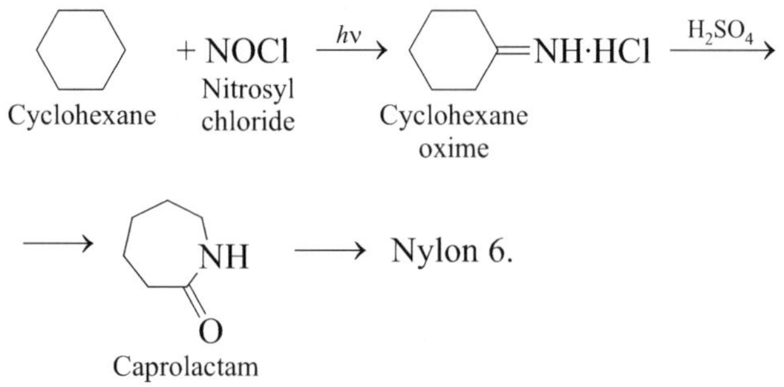

Cyclohexane + NOCl (Nitrosyl chloride) →hv→ Cyclohexane oxime (=NH·HCl) →H$_2$SO$_4$→

→ Caprolactam (NH, O) → Nylon 6.

2.6.5.8 Paterno-Büchi Reaction

Photochemical cycloaddition of carbonyl compounds with olefins gives oxetanes (four-membered ether rings). The reaction is known as Paterno-Büchi reaction.

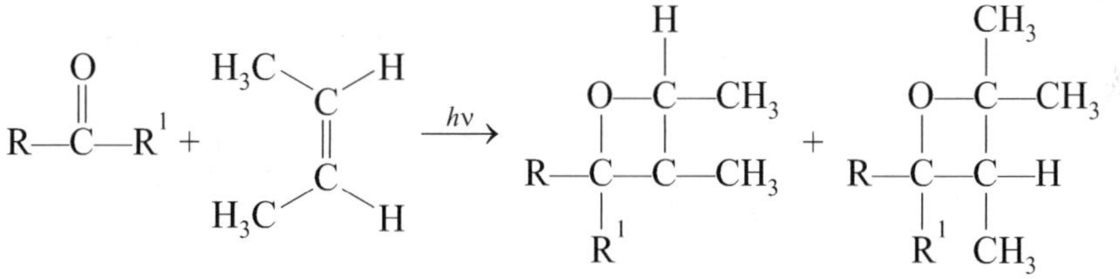

The Paterno-Büchi reaction usually occurs by the cycloaddition of the triplet state of the carbonyl compound with the ground state of an alkene. An interesting reaction is the photo addition of butyraldehyde with 2-methyl-2-butene to yield a mixture of 2,3,3-trimethyl-4-propyloxetane and 2,2,3-trimethyl-4-propyloxetane.

$$CH_3CH_2CH_2CHO \ + \ CH_3-C=CH-CH_3$$

Butyraldehyde

CH$_3$ (on the central carbon)

2-Methyl-2-butene

2,3,3-Trimethyl-4-propyloxetane   +   2,2,3-Trimethyl-4-propyloxetane

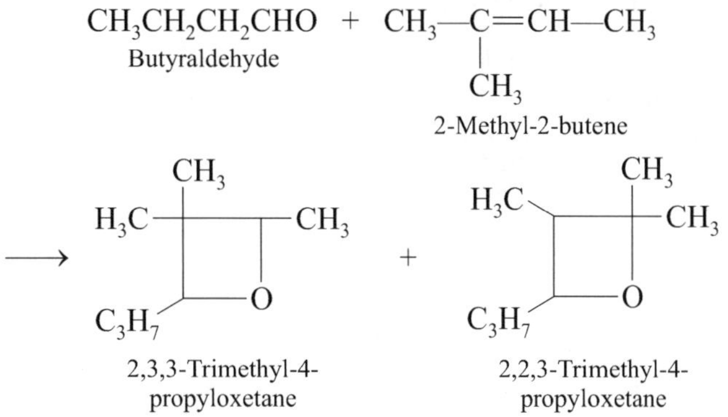

The oxetane ring is formed in two steps. The carbonyl compound (triplet state) adds through its oxygen atom to give the more stable diradicals. In the second step, the spin inversion occurs with simultaneous bond formation to give oxetane.

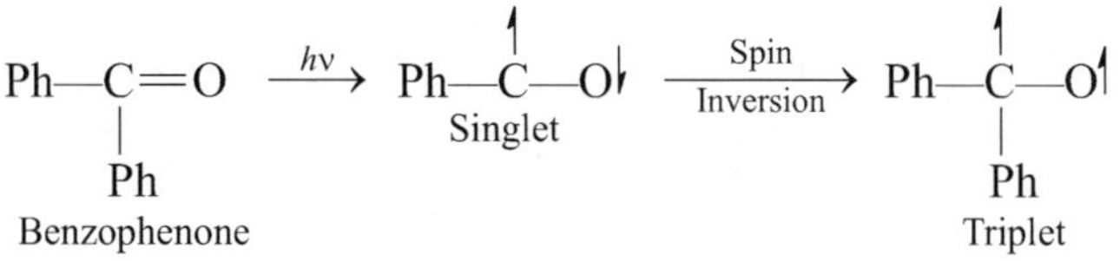

Ph—C=O (Ph) Benzophenone →hv→ Ph—C—O· Singlet →Spin Inversion→ Ph—C—O· (Ph) Triplet

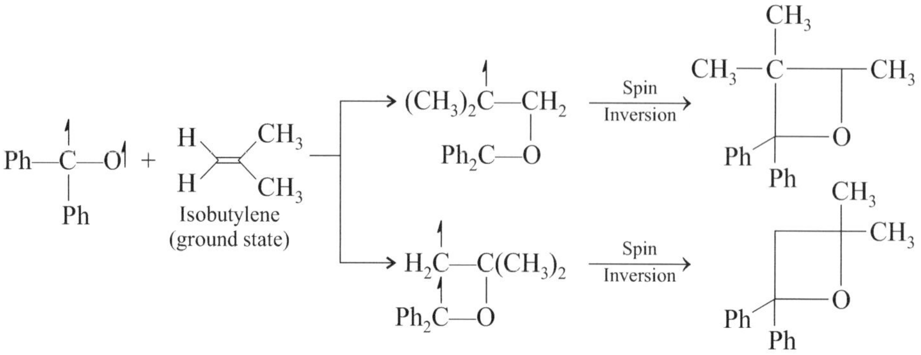

The photocycloaddition of benzophenone with cis- and trans-2-butene gives the same mixture of cis- and trans- oxetanes showing that the reaction is not stereospectic. The lack of stereochemical discrimination shows that the reaction is not concerted and that the ring is formed in two stages as shown below:

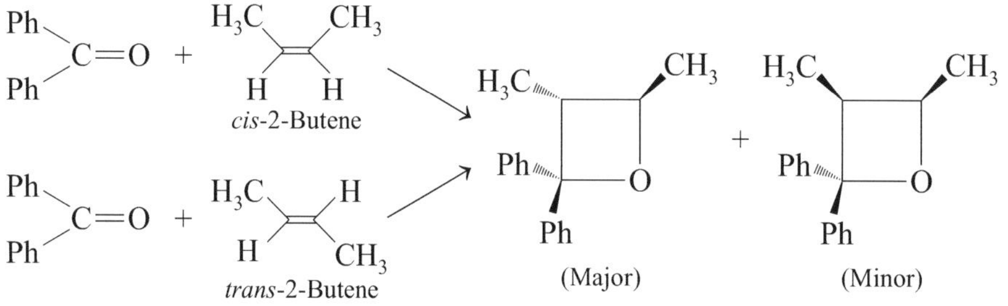

2.6.5.9 Formation of Vitamin D₂ from Ergosterol

Irridation of the steroid ergosterol with UV light gives *cis*-tachysterol (pre vitamin D₂) by a ring opening process. The *cis*- tachysterol thus obtained is thermally isomerised to vitamin D₂ (calciferol).

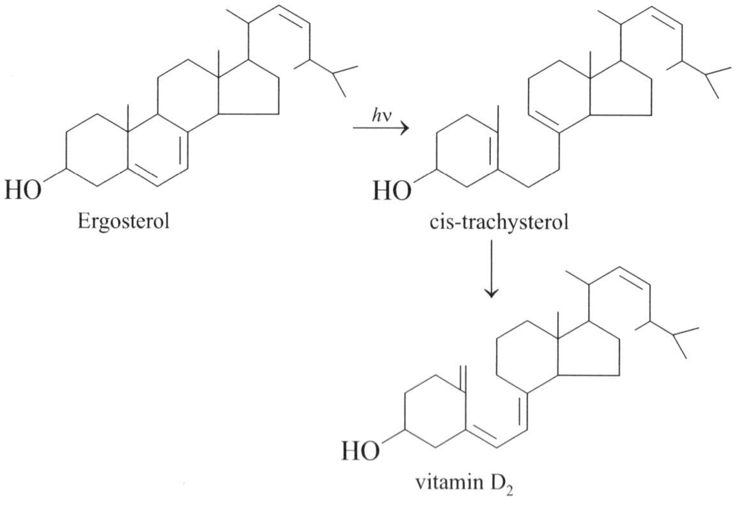

Similarly, vitamin D_3 is formed from 7-dehydro cholesterol. Milk and other foods are irradiated with light in order to increase the vitamin D content.

Out of all the reactions discussed, photochemical reactions are the best for organic synthesis because these reactions produce no byproducts.

2.6.6 Oxidations

A large number of reactions in organic synthesis involve oxidation. In general, oxidation of an organic susbstrate involve addition of oxygen or removal of hydrogen. Some examples are:

$$R\text{—}CHO \xrightarrow{[O]} R\ COOH$$

$$RCH{=}CH\text{—}R \xrightarrow{RCO_3H} R\text{—}CH\text{—}CH\text{—}R$$

$$R_2\text{—}CHOH \xrightarrow{[O]} R_2CO$$

Addition of oxygen

(Removal of hydrogen)

In an oxidation reaction, the reagent gets reduced. For example, during oxidation of secondary alcohol to ketone with $KMnO_4$ Mn(VII) is converted into Mn(II) or Mn(IV).

$$5R_2CHOH + 2MnO_4 + 3H_2SO_4 \longrightarrow 5R_2CO + 2MnSO_4 + K_2SO_4 + 8H_2O$$

Similarly, in oxidation using chromium oxide, the hexavalent chromium (VI) is reduced to chromium (III).

Oxidation may also involved removal of electron from the substrate. Loss of electron is associated with change in oxidation state of an atom (in the functional group). This loss of electrons can be determined by finding the oxidation number. A rule used for the determining of oxidation number is used only for atoms which are commonly encountered (H, O, C or halogens) in simple molecules. As per convention, oxidation number for hydrogen is –1; oxidation number for carbon (bonded to the same atom) is O and oxidation number for carbon bonded to a hetero atom (oxygen or halogen) is + 1. Only the atoms that are modified or changed during the oxidation are examined. By comparing the sum of oxidation numbers of various atoms in the starting material and in the product formed the change in oxidation number can be calculated.

As an example oxidation of ethanol gives acetaldehyde, which is further oxidised to acetic acid.

$$CH_3CH_2OH \xrightarrow{2e^- \text{ lost}} CH_3CHO \xrightarrow{2e^- \text{ lost}} CH_3\text{—}\overset{\overset{\displaystyle O}{\|}}{C}\text{—}OH$$
$$\quad -1 \qquad\qquad\qquad -1 \qquad\qquad\quad -3$$

In the above example, in ethanol the carbon bearing the hydroxyl group has an oxidation number of -1 (C is 0, H is -1, O is $+1$). The oxidation number of that carbon in acetaldehyde is $+1$ (C is O, H is -1, O is $+1$) and the oxygen atom from the second bond to the carbonyl is $+1$. The change in oxidation number is $-1 \longrightarrow +1$, for a net loss of two electrons. So this is an oxidation reaction. In a similar way, the carbon of interest in acetic acid has an oxidation number of $+3$ (C is 0; the three bonds to oxygen total $+3$). Thus, conversion of a acetaldehyde to acetic acid involves loss of two additional electrons ($+1 \longrightarrow +3$) and is an oxidation reaction.

For oxidation a number of oxidising agents are available. The choice of the oxidising agent depends on the compound to be oxidised. Following are given oxidation of some families of compounds like hydrocarbons, alcohols and aldehydes and ketones.

2.6.6.1 Oxidation of Hydrocarbons

2.6.6.1.1 Oxidation of Alkanes

Alkanes are known to be inert to many chemical reagents and are uneffected by oxidising agents under normal conditions.

Alkanes can be oxidised under catalytic conditions. Two such process of commercial importance are manufacture of methyl alcohol and formaldehyde.

$$\underset{\text{Methane}}{CH_4} + \frac{1}{2}O_2 \xrightarrow[\text{100 atm., 260°C}]{\text{Cu catalyst}} \underset{\text{Methyl alcohol}}{CH_3OH}$$

$$\underset{\text{Methane}}{CH_4} + O_2 \xrightarrow[\text{MoO, 400°C, 200 atm.}]{\text{controlled } oxidation} \underset{\text{Formal aldehyde}}{HCHO + H_2O}$$

2.6.6.1.2 Oxidation of Alkenes

The presence of double bond in alkenes makes them more reactive.

(i) Oxidation with potassium permanganate: The alkenes on oxidation with dilute $KMnO_4$ solution in alkaline medium give diol (glycol).

This reaction forms the basis of **Baeyer's test**, which is used for the detection of unsaturation.

Hot $KMnO_4$ in alkaline medium cleaves the double bond of the alkene. Thus, oxidation of *cis*- or *trans*- 2-butene with alkaline $KMnO_4$ (refluxing) gives acetic acid.

$$CH_3CH\!=\!CHCH_3 \xrightarrow[\text{heat}]{\text{KMnO}_4,\ H_2O,\ ^-OH} 2CH_3C\!\!\underset{O^-}{\overset{O}{\diagup}} \xrightarrow{H^+} 2CH_3\overset{O}{\overset{\|}{C}}\!\!-\!OH$$

2-Butene   Acetate ion   Acetice acid
(cis-or trans)

Terminal CH_2 group of 1-alkene on oxidation with hot $KMnO_4$ give keto compounds.

$$CH_3CH_2\overset{\overset{\displaystyle CH_3}{|}}{C}\!=\!CH_2 \xrightarrow[\text{2) } H_3O^+]{\text{1) KMnO}_4,\ ^-OH,\ \text{heat}} CH_3CH_2\overset{\overset{\displaystyle CH_3}{|}}{C}\!=\!O + O\!=\!C\!=\!O + H_2O$$

2-methyl-1-butene   Ethyl methyl
                                        ketone

Much better yields are obtained in case the oxidation with aqueous $KMnO_4$ is carried in presence of a phase transfer catalyst. Thus,

$$CH_3(CH_2)_5CH\!=\!CH_2 \xrightarrow[\substack{CH_3(CH_2)_{15}\overset{+}{N}(CH_3)_3\overset{-}{C}l\\ \text{(PTC)}}]{aq.\ \text{KMnO}_4} CH_3(CH_2)_4CH_2COOH$$

1–Octene   Heptanoic acid

(*ii*) **Oxidation with chromium reagents:** Oxidation of alkenes with chromyl chloride at low temperature gives carbonyl compounds.

$$CH_3\!-\!\overset{\overset{\displaystyle CH_3}{|}}{\underset{\underset{\displaystyle CH_3}{|}}{C}}\!-\!CH_2\!-\!\overset{\overset{\displaystyle CH_3}{|}}{C}\!=\!CH_2 \xrightarrow[\substack{CH_2Cl_2\\ 0-5°C}]{CrO_2Cl_2} CH_3\!-\!\overset{\overset{\displaystyle CH_3}{|}}{\underset{\underset{\displaystyle CH_3}{|}}{C}}\!-\!CH_2\!-\!C\!\!\!\overset{\overset{\displaystyle CH_3}{|}}{\underset{}{\diagdown}}\!\!\!\overset{\overset{\displaystyle H}{|}}{\underset{}{C}}\!-\!H$$

2, 4, 4-Trimethyl-1
-pentene

$$\Big\downarrow Zn/H_2O$$

$$CH_3\!-\!\overset{\overset{\displaystyle CH_3}{|}}{\underset{\underset{\displaystyle CH_3}{|}}{C}}\!-\!CH_2\!-\!\overset{\overset{\displaystyle CH_3}{|}}{CH}\!-\!CHO$$

2, 4, 4-Trimethyl-
pentanal (75%)

Internal alkenes on oxidation with CrO_3/H_2SO_4 give ketones

$$CH_2\!=\!CHCH_3 \xrightarrow[\substack{Hg(OCOEt)_2,\ Me_2CO\\ 25°C,\ 4\ hr.}]{CrO_3,\ H_2O,\ H_2SO_4} CH_3\!-\!\overset{\overset{\displaystyle}{}}{\underset{\underset{\displaystyle O}{\|}}{C}}\!-\!CH_3$$

Propene   Acetone
                                        (80%)

(*iii*) **Oxidation with peracids:** Alkenes on reaction with peracids give the corresponding epoxide. The reaction takes place by syn addition of epoxide to the double bond.

$$C=C \longrightarrow C-C \quad + \quad R-C-OH$$

Peracid → Epoxide + carboxylic acid

The epoxidation with peracids is stereoselective. Thus, cis and trans alkene give only the cis and trans epoxides respectively. The expoxidation, in fact takes place by a concerted mechanism (onc-step mechanism).

cis-olefin $\xrightarrow{C_6H_5CO_3H}$ cis-Epoxide

trans-olifin $\xrightarrow{C_6H_5CO_3H}$ trans-Epoxide

Besides peracids, epoxidation of alkenes can also be effected by dimethyl dioxirane and H_2O_2.

The process of epoxodation followed by hydrolysis of the epoxide is a convenient method for trans hydroxylation of alkenes.

Alkene $\xrightarrow{RCO_3H}$ Epoxide $\xrightarrow{H_2O}$ $\longrightarrow$ $\longrightarrow$ trans diol

cis- Hydroxylation of alkenes can however be effected by oxidation with $KMnO_4$ or chromium oxide followed by hydrolysis.

Alkene $\xrightarrow{2H_2O}$ cis-Diol $+ H_2MnO_4$

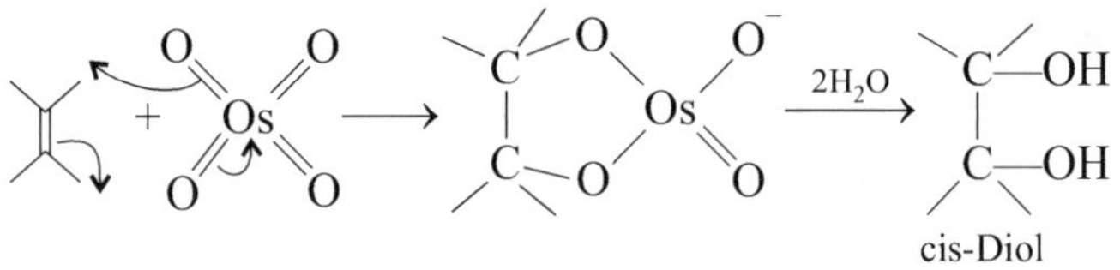
2H$_2$O
cis-Diol

(*iv*) **Hydoboration-Oxidation:** Treatment of alkenes with borane followed by oxidation gives anti-Markovnikoff addition product. However Markovnikoff adduct can be obtained by hydration of alkene with dilute H_2SO_4.

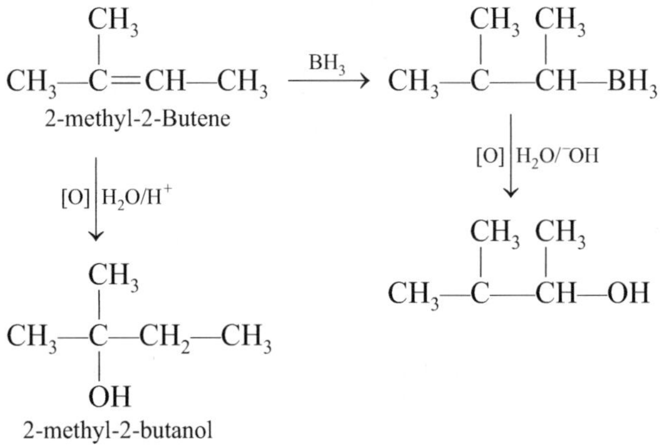
BH$_3$
2-methyl-2-Butene
[O] H$_2$O/H$^+$
[O] H$_2$O/$^-$OH
2-methyl-2-butanol

(*v*) **Ozonolysis:** Reaction of alkene with ozone gives an adduct (called ozonide) which on treatment with a suitable reducing agent gives two carbonyl compounds

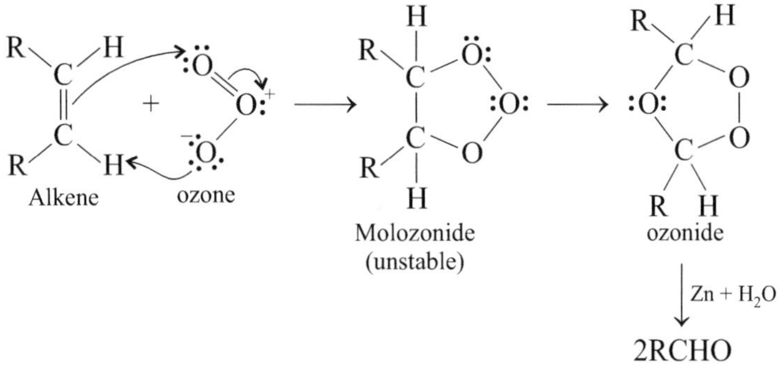
Alkene
ozone
Molozonide
(unstable)
ozonide
Zn + H$_2$O
2RCHO

A typical example is given below:

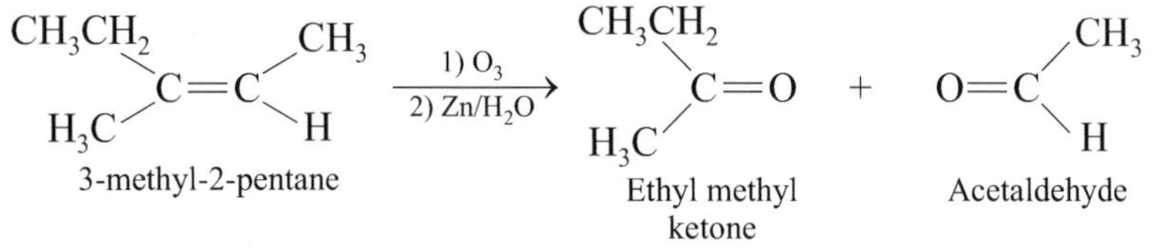
1) O$_3$
2) Zn/H$_2$O
3-methyl-2-pentane
Ethyl methyl ketone
Acetaldehyde

2.6.6.1.3 Oxidation of Alkynes

Alkynes undergo oxidative cleavage with ozone or with alkaline $KMnO_4$ to give carboxylic acids.

$$R—C\equiv C—R' \xrightarrow[\text{(2) HOAC}]{\text{(1) O}_3} RCO_2H + R'CO_2H$$

$$R—C\equiv C—R' \xrightarrow[\text{(2) H}^+]{\text{(1) KMnO}_4\text{/OH}^-} RCO_2H + R'CO_2H$$

Terminal alkynes undergo oxidative coupling to give diacetylenes by oxygen in presence of cuprous salts

$$C_6H_5C\equiv CH \xrightarrow[\text{40 min}]{\text{O}_2\text{CuCl, 30 – 40°C}} C_6H_5C\equiv C—C\equiv C—C_6H_5$$

Alkynes on hydroboration followed by oxidation with H_2O_2 give aldehydes or ketones depending on the structures of the aldehyde

$$CH_3CH_2C\equiv CH \xrightarrow[\substack{\text{Disiamyl}\\ \text{borane}}]{(C_5H_{11})_2BH} CH_3CH_2CH=CHB(C_5H_{11})_2$$

1-Butyne
(Terminal alkyne)

$$\xrightarrow{[O] \downarrow H_2O_2}$$

$$CH_3CH_2CH_2CHO \rightleftharpoons CH_3CH_2CH=CHOH$$

Butanal

2.6.6.1.4 Oxidation of Aromatic Hydrocarbons

Benzene being an exceptionally stable compound is not affected by usual oxidizing agents. However, it undergoes catalytic oxidation by air at 400–450°C in presence of V_2O_5 to give maleic anhydride, which on boiling with water gives maleic acid. This is a commercial method for the preparation of maleic acid.

Benzene
Maleic anhydride
Maleic acid

Condensed aromatic compounds such as naphthalene, anthracene and phenanthracene on oxidation with a suitable oxidising agent give 1, 4-napthoquinone, anthraquinone and 9, 10-phenanthrenquinone respectively.

Naphthaline, R = H
2, 3-Dimethylnaphthaline, R = CH₃

1, 4-Naphthoquinone (32–35%)
2, 3-Dimethyl-1, 4-naphthoquinone
(60–80%)

Anthracene

$$\xrightarrow[\text{Bu}_4\text{NHSO}_3]{\text{Na}_2\text{Cr}_3\text{O}_3/\text{H}_2\text{SO}_4}$$
70°C, 2 min

Anthraquinone
(80–90%)

Phenanthracene

$$\xrightarrow[\text{H}_2\text{O, 1 hr.}]{\text{K}_2\text{Cr}_2\text{O}_7/\text{H}_2\text{SO}_4}$$

9, 10-Phenanthrenequinone
(80%)

Catalytic oxidation of naphthalene or *o*-xylene gives phthalic anhydride. It is a industrial method for the manufacture of phthalic anhydride.

Naphthaline $+ 4\frac{1}{2}O_2$ (air) $\xrightarrow[500°C]{V_2O_5}$ Phthalic anhydride $\xleftarrow[500°C]{[O] \atop V_2O_5}$ *o*-Xylene

In case, the benzene nucleus has a side chain, oxidation with alkaline $KMnO_4$ gives benzoic acid in 40-50% yield. However, oxidation with $KMnO_4$ in presence of a phase transfer catalyst (PTC) gives 80% yield in much shorter time.

Toluene, R = CH_3
Ethylbenzene, R = CH_2CH_3

Isopropylbenzone, R = $-CH(CH_3)_2$

$\xrightarrow{[O]}$

Benzoic acid
alk. $KMnO_7$ reflux 8-10 hr,
yield 40–50%
cetylammonium chloride
(PTC) yield 80%

Toluene can be oxidised to benzaldehyde by chromyl chloride (**Etard Reaction**).

$$C_6H_5CH_3 \xrightarrow[\text{(2) } H_3O^+]{\text{(1) Cr O}_2\text{Cl}_2} C_6H_6CHO$$
Tolume   Benzaldehyde (70 - 80%)

Benzene ring in toluene, chlorobenzene and styrene on oxidation with enzyme of *Pseudomonas putida* undergoes stereospecific *syn*-hydroxylation at positions 2 and 3 to give the corresponding *cis*-2, 3-dihydrocyclohexa-4, 6-dienes.

R = CH_3, Cl or $-CH=CH_2$

$$\xrightarrow[\text{air, 20°C, 10–20 hr.}]{\text{Pseudomonas putida}}$$

cis-2, 3-dihydrocyclohexa
-4, 6-dienes (6–10%)

2.6.6.2 Oxidation of Alcohols

(*i*) **Oxidation of primary alcohols:** Primary alcohols on oxidation give aldehydes which is subsequently oxidised to carboxylic acid.

$$R\text{—}CH_2CH_2OH \xrightarrow{[O]} [RCH_2CHO] \xrightarrow{[O]} RCH_2COOH$$

1° Alcohol   Intermediate aldehyde   Carboxylic acid

The intermediate aldehyde can be obtained in good yield depending on the oxidizing agent used.

Oxidation of primary alcohols directly to carboxylic acids can be effected by $KMnO_4$.

Primary alcohols can be oxidised to the aldehydes with $Na_2Cr_2O_7$ or $K_2Cr_2O_7$ in dilute H_2SO_4.

Satisfactory yield of the aldehyde can be obtained by continuously distilling off the formed aldehyde.

$$3RCH_2OH + Cr_2O_7^{2-} + 4H_2SO_4 \longrightarrow 3RCHO + 7H_2O + 2Cr_3^{+} + 4SO_4^{2-}$$

Oxidation of primary alcohols to aldehydes can also be effected by $Na_2Cr_2O_7$ in presence of PTC or with chromyl chloride.

$$C_7H_{15}CH_2OH \xrightarrow[\substack{Bu_4\,NHSO_4,\,CH_2Cl_2 \\ RT,\,1\,min.}]{\substack{Na_2\,Cr_2O_7,\,3\,M\,H_2SO_4}} C_7H_{15}CHO$$
95%

$$\xrightarrow[\text{RT, 5 hr.}]{CrO_2Cl_2 \text{ in } CH_2Cl_2} C_7H_{15}CHO$$
94%

Good yields of the aldehydes can also be obtained by the oxidation of primary alcohols with a solution of Cr_2O_3 in dilute H_2SO_4 (**Jones reagent**), or **collins reagent** (prepared by adding in small portions chromium trioxide to pyridine) or by pyridinium chlorochromate (PCC, C_5H_5NH CrO_3Cl), prepared by adding pyridine to a solution of chromium trioxide in 6M HCl) or by MnO_2.

$$\xrightarrow[\text{Jones reagent}]{[O]}$$

85%

$$C_6H_5CH{=}CHCH_2OH \xrightarrow[RT]{CrO_3,\,C_5H_5N} C_6H_5CH{=}CH\text{—}CHO$$
85%

$$C_5H_{11}C{\equiv}CCH_2OH \xrightarrow[20°C,\,1-2\,hr]{C_5H_5NHCrO_3Cl} C_5H_{11}C{\equiv}C\text{—}CHO$$
84%

$$C_6H_5CH_2OH \xrightarrow[\text{chloroform}]{MnO_2,\,RT\,24\,hr} C_6H_5CHO$$
89%

(*ii*) **Oxidation of secondary alcohols:** Secondary alcohols on oxidation with $KMnO_4$ (in acidic or alkaline conditions), $Na_2Cr_2O_7/H_2SO_4$, Cr_2O_3, Jones reagent, collins reagent, pyridinium chlorochromate (PCC) or pyridinium dichromate give ketones. A good useful reagent for the oxidation of 1°

and 2° alcohols to aldehydes and ketones in good yield is **Des-Martin reagent**, which is prepared from *o*-iodobenzoic acid as given below.

A convenient method for the oxidation of alcohols to aldehydes or ketones is achieved by *tetra-n*-propylammonium perruthenate (TPAP). This oxidant has been used to effect oxidation in complex molecules without disturbing other functional groups. Aldehydes bearing a labile α-stereogenic centre can be prepared without any change of stereochemistry. An advantage of this oxidant is that it can be used as a catalyst (5 mol %) in conjuction with an excess of N-methylmorpholine N-oxide (NMO) (which reoxidises the ruthenium in situ). This reagent combination is useful in large scale synthesis as long as the reaction is carried out under anhydrous conditions.

Tertiary alcohols are resistant to oxidation.

2.6.6.3 Oxidation of Phenols

Oxidation of phenols with alkaline potassium persulphate give the corresponding para hydroxy phenol. The reaction is known as **Elbs hydroxylation**

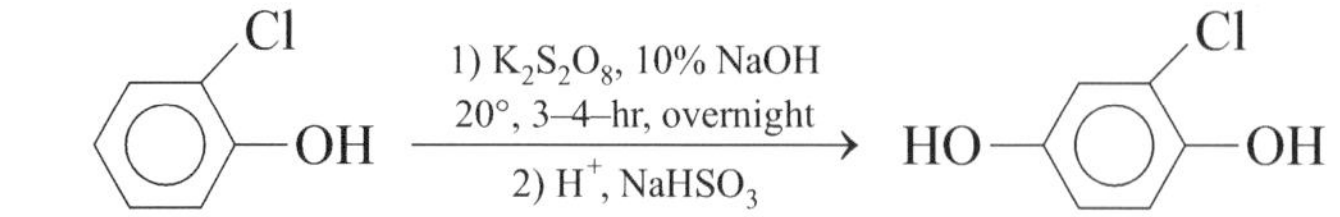

o-Chlorophenol

2, 5, Dihydroxy-chlorobenzene (62%)

1) $K_2S_2O_8$, 10% NaOH
20°, 3–4–hr, overnight
2) H^+, $NaHSO_3$

A common oxidation of phenol is its conversion to the corresponding ortho or para-quinones by mercuric oxide or mercuric trifluoroacetate, lead dioxide, chromium trioxide, DDQ or Fermy's salt.

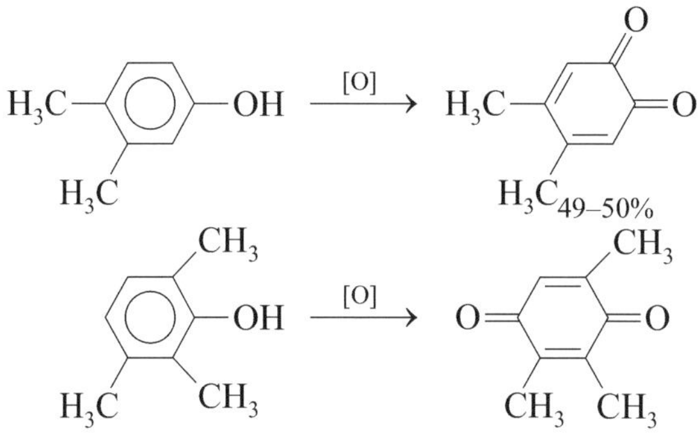

2, 3, 6-Trimethyl phenol

2, 3, 6-Trimethyl benzoquinone (77–79%)

Hydroquinone on oxidation with $Na_2Cr_2O_7/H_2SO_4$ or $KBrO_3$ gives p-benzoquinone is 75-80% or 90% yield.

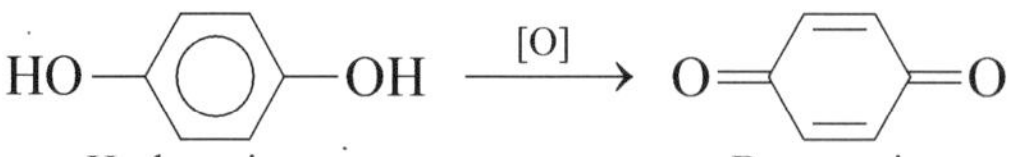

Hydroquinone

p-Benzoquinone

2.6.6.4 Oxidation of Aldehydes

Both aliphatic and aromatic aldehydes on oxidation with $KMnO_4$ give the corresponding carboxylic acids. Much better yield of carboxylic acid is obtained if oxidation is carried out with tetrabutylammonium permanganate (Bu_4NMnO_4, C_5H_5N) or benzyltriethyl ammonium permanganate ($PhCH_2NEt_3MnO_4$, CH_3COOH, CH_2Cl_2, 1.25 hr)

$$RCHO \xrightarrow{[O]} RCOOH$$

Aldehydes can also be oxidised by **Tollens reagent** which is prepared by adding 1-2 drop of 5% NaOH solution to a solution of $AgNO_3$ in H_2O till a turbidity remains. This is followed by addition of NH_4OH to get a clear solution.

$$RCHO + 2[Ag(NH_3)_2]OH \xrightarrow{[O]} RCOONH_4 + 3NH_3 + H_2O + 2Ag$$

Tollens reagents   Silver mirror

$$\downarrow H^-$$ or

                                                         grey-black ppt.

$$RCOOH$$

This is used as a test for CHO group.

Aldehydes can also be oxidised to the corresponding carboxylic acids by heating with **Fechling solution** [consisting of two solutions: Fechling A (obtained by dissolving crystalline $CuSO_4$ in water containing few drops of H_2SO_4) and Fechling B (obtained by dissolving sodium potassium tartarate (Rochell salt) in NaOH solution) Before use both Fechling A and Fechling B solutions are mixed in equal amounts]. In this reaction aldehydes are oxidised to carboxylic acids and Cu^{2+} is reduced to Cu^{+1} as red precipitate of Cu_2O.

This test is also useful for detecting the presence of CHO group.

Aldehydes can also be oxidised with sodium hypohalite (a solution of iodine in potassium iodine in presence of NaOH solution)

$$H_3CCHO \xrightarrow[\text{NaOH}]{I_2/KI} I_3CCHO \xrightarrow{-OH} CHI_3 + HCOONa.$$

ioloform

This reaction is called **Haloform reaction**. The iodoform obtained has a pleasant odour. This reaction is also useful for detecting the presence of CHO group in aliphatic aldehydes.

The mechanism of the Haloform reaction is given below:

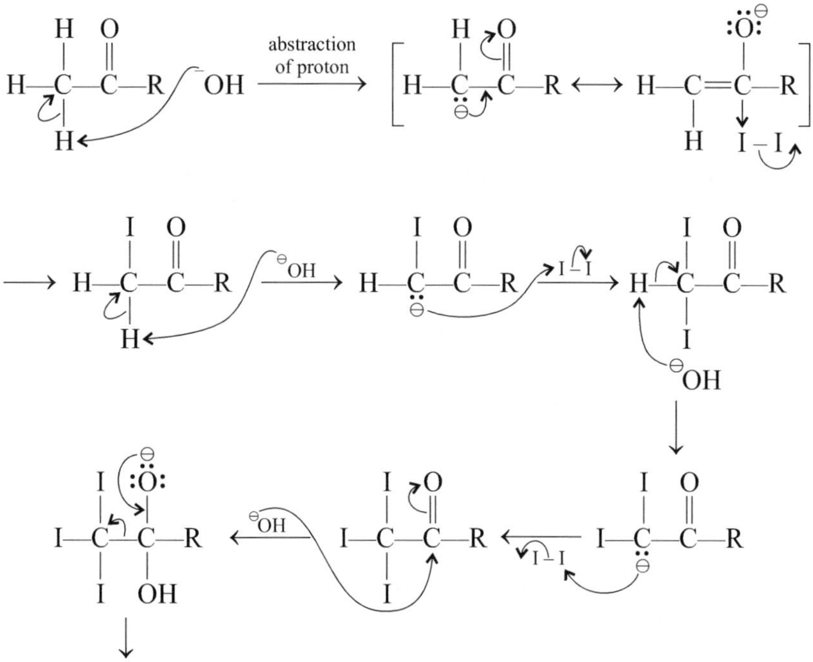

(A)

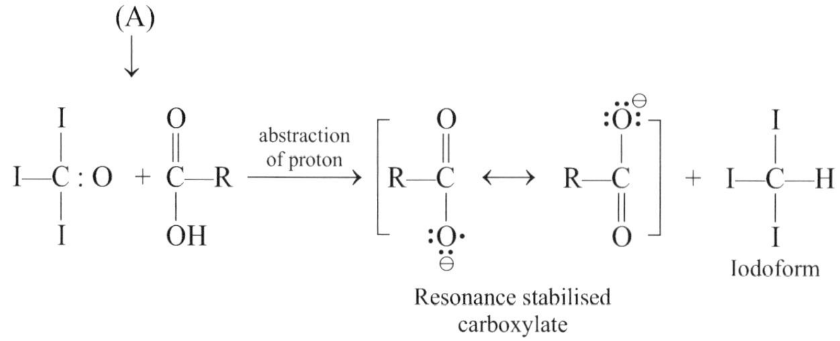

(A)

$$I\!-\!\overset{\displaystyle I}{\underset{\displaystyle I}{C}}\!:\!O \;+\; \overset{\displaystyle O}{\underset{\displaystyle OH}{C}}\!-\!R \;\xrightarrow[\text{of proton}]{\text{abstraction}}\; \left[\underset{\displaystyle :\!\overset{..}{O}\!\cdot}{R\!-\!\overset{\displaystyle O}{C}} \longleftrightarrow \underset{\displaystyle O}{R\!-\!\overset{:\ddot{O}:^{\ominus}}{C}} \right] \;+\; I\!-\!\overset{\displaystyle I}{\underset{\displaystyle I}{C}}\!-\!H$$

Resonance stabilised
carboxylate

Iodoform

Aromatic aldehydes, especially those containing hydroxyl group are oxidised to the corresponding phenols (CHO $\longrightarrow$ OH) with alkaline H_2O_2. This reaction is known as **Dakins oxidation**.

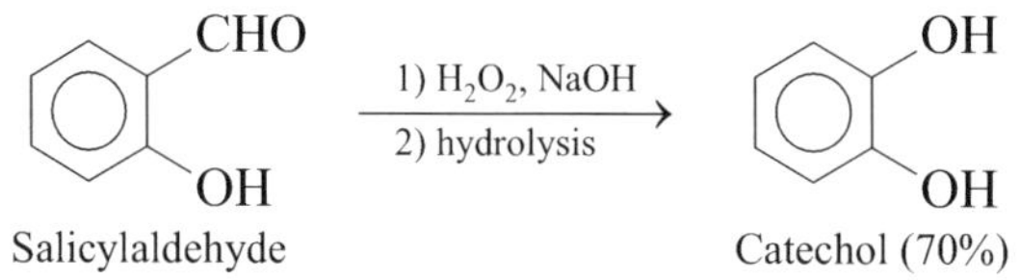

$$\text{Salicylaldehyde} \xrightarrow[\text{2) hydrolysis}]{\text{1) } H_2O_2, \text{ NaOH}} \text{Catechol (70\%)}$$

The mechanism of the reaction is given below:

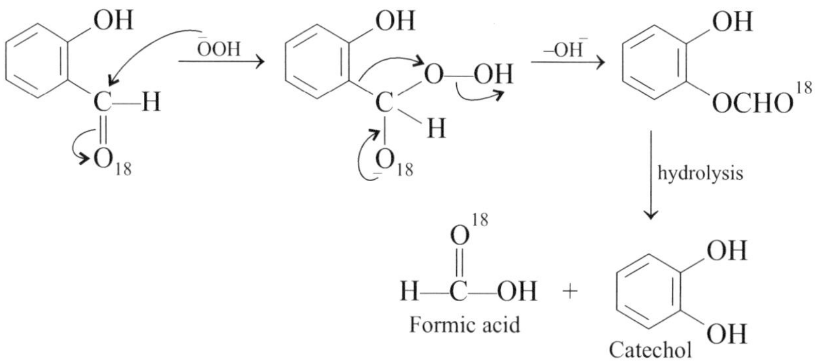

[Reaction mechanism scheme: salicylaldehyde with $\overline{O}OH$ attack forming intermediate with O—OH, then $-OH^-$ giving aryl $OCHO^{18}$ ester, which on hydrolysis gives Formic acid $H-\overset{O^{18}}{C}-OH$ + Catechol]

Formic acid

Catechol

In place of alkaline H_2O_2, oxidation can also be effected with peroxy acids like *m*-chloroperbenzoic acid. This reaction is known as Baeyer-Villiger reaction.

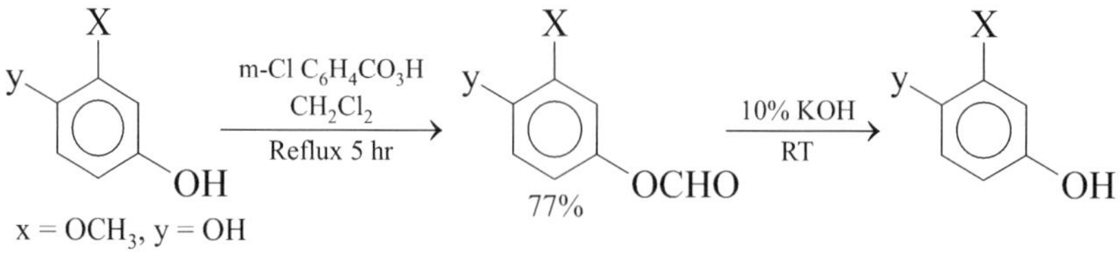

$$\xrightarrow[\text{Reflux 5 hr}]{\substack{\text{m-Cl } C_6H_4CO_3H \\ CH_2Cl_2}} \; (77\%) \; \xrightarrow[\text{RT}]{10\% \text{ KOH}}$$

x = OCH₃, y = OH

For mechanism of the reaction, see page 154.

2.6.6.5 Oxidation of Ketones

The ketones (compared to aldehydes) are resistant to oxidation. This is because ketones do not contain any H attached to the carbonyl group. However, under vigorous conditions, the cleavage of C—C bond occurs giving a mixture of acids

$$CH_3CH_2\overset{\overset{\displaystyle O}{\|}}{C}-CH_2CH_3 \xrightarrow[\Delta]{KMnO_4/H^+} CH_3CH_2\overset{\overset{\displaystyle O}{\|}}{C}-OH + CH_3COOH$$

3-Pentanone

Proponic acid

Acetic acid

In case of unsymmetrical ketones, the cleavage occurs in a way so that the CO remains with the smaller alkyl group.

$$CH_3\overset{\overset{\displaystyle O}{\|}}{C}CH_2CH_2CH_3 \xrightarrow[\Delta]{KMnO_4/H^+} CH_3-\overset{\overset{\displaystyle O}{\|}}{C}-OH + CH_3CH_2COOH$$

2-Pentanone

Acetic acid

Proponic acid

Methyl ketones undergo oxidation with sodium hypohalite (Haloform reaction).

$$Ph\overset{\overset{\displaystyle O}{\|}}{C}CH_3 + 4NaOH + 3I_2 \longrightarrow Ph\overset{\overset{\displaystyle O}{\|}}{C}-\overset{-}{O}\overset{+}{Na} + CHI_3 + 3NaI + 3H_2O$$

Acetophenone

Sod. salt of benzoic acid

Iodifon

For mechanism of haloform reaction see page 175.

Both aliphatic and aromatic ketones can be oxidised with per acids to give esters, which an hydrolysis give carboxylic acid and alcohol (Baeyer-villiger oxidation).

Cyclic ketones such as cyclohexanone on Baeyer-villiger oxidation give ε-carprolactone.

cyclohexane

20% RCO_3H, HCO_2Et

6 hr

70–75%
ε-caprolactone

2.6.6.6 Oxidation of Amines

Primary amines (both aliphatic and aromatic) on treatment with dimethyl dioxirane (prepared by reacting acetone with oxone) give the corresponding nitrocompounds.

$$\overset{CH_3}{\underset{CH_3}{\diagdown}}C=O \xrightarrow{oxone} \overset{CH_3}{\underset{CH_3}{\diagdown}}C\overset{O}{\underset{O}{\diagup}}$$

[Oxone is $2KHSO_4 \cdot KHSO_4 \cdot K_2SO_4$]

Some examples are given below

Propylamine   Nitropropane
                                                (80–85%)

Cyclohexylamine   Nitrocyclohexane

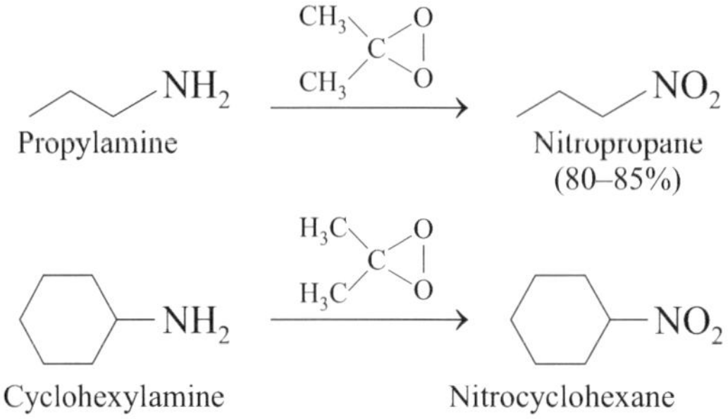

However secondary amines on reaction with dimethyldioxirane give the corresponding hydroxylamine.

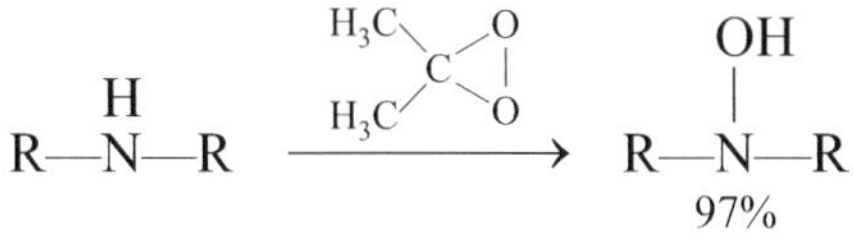

97%

Tertiary amines, since they do not contain N—H bonds are not oxidised. However, these react with per acids to give the corresponding N-oxides.

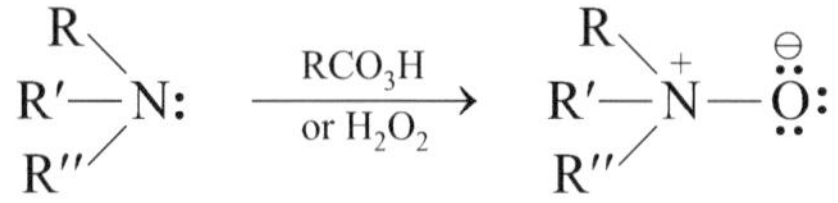

2.6.7 Reductions

Like oxidation, reduction also plays an important role in synthetic organic chemistry. Only a few organic synthesis do not involve reduction at some stage.

Reduction is defined as removal of oxygen (or decrease in oxygen content), addition of hydrogen or addition of electrons to an organic substrate. Some examples are given below:

$$R-\overset{\overset{\displaystyle O}{\|}}{C}-OH \xrightarrow[\text{reduction}]{[H]} R-\overset{\overset{\displaystyle O}{\|}}{C}H \quad \text{(decrease in oxygen content)}$$

Carboxylic acid

$$R-\overset{\overset{\displaystyle H}{|}}{C}=O \xrightarrow[\text{reduction}]{[H]} R-CH_2OH \quad \text{(addition of hydrogen)}$$

aldehyde   (increase in hydrogen content)

$$CH_3CH_2COCH_3 \xrightarrow[\text{reduction}]{[H]} CH_3CH_2CHOHCH_3 \quad \text{(addition of electrons)}$$

2-Butannone   (Gain 2e⁻)

In order to decide whether a particular reaction is reduction or not can be decided as follows. Reduction involves conversion of an atom from a higher oxidation state to a lower oxidation state. For assigning the oxidation number, the following procedure in used for commonly encountered atoms like H, O, C, N and halogen. As per as the conversion, the oxidation number for hydrogen is – 1: oxidation number for carbon (bonded to the same atom) is O; oxidation number for carbon bonded to a hetero atom (O or halogen) is +1.

The oxidation number of only those atoms that are changed in the redution are examined. If the sum of oxidation number of various atoms in the substrate and the product are compared, the change in oxidation number can be calculated. From the change it is possible to find out whether a particular reaction is oxidation or reduction. As an example, in the redution of propanal to propanol, the oxidation state of C in CHO changes from + 1 to – 1 for C in CH_2OH.

$$CH_3CH_2CHO \xrightarrow{\text{gain of } 2e-} CH_3CH_2CH_2OH$$
$$\underset{+1}{} \qquad\qquad\qquad\qquad\qquad \underset{-1}{}$$

In the example cite above, in propanol the carbon atom of the aldehyde group has an oxidation number + 1 (C is 0, H is – 1, O is + 1 and O from second bond to the carbonyl is + 1). However in the case of the product, propyl alcohol, the oxidation number of C in CH_2OH is – 1 (C is 0, H is – 1, H is – 1 and O is + 1). It is seen that the change in oxidation number is + 1 → – 1 for a net gain of 2 electrons. So the reacts is a reduction reaction.

The product in a reduction depends on the nature of the substrate, the nature of the reducing agent and reaction conditions. The method depends on the selectivity and the stereochemistry of the desired product.

Reductions an be effected chemically, or catalytic hydrogenations. Besides certain enzymes can also be used to effect reductions (enzymatic reductions).

Following are given reduction of some families of compounds like unsaturated hydrocarbons, aldehydes, ketones, carboxylic acid and its derivatives, and aromatic nitro compounds.

2.6.7.1 Reduction of Unsaturated Compounds

The unsaturated compound include alkenes, alkynes and aromatic hydrocarbon.

2.6.7.2 Reduction of Alkenes

Reduction of alkenes can be effected in presence of finely divided metals like Pt, Ni, Pd, Rh etc to give a syn addition product. In this case both H atoms are added from the same side of the molecule.

Hindered alkenes are difficult to reduce and vigorous conditions are required. The choice of the reducing agent depends on the nature of the other functional groups present and also the selectively required. In some cases, Raney Ni is used. An example is given below:

$$C_6H_5CH=CHCH_2OH \xrightarrow[\text{EtOH}]{\text{Raney Nickel/H}_2} C_6H_5CH_2CH_2CH_2OH$$

cinnamyl alcohol   3-phenyl propanol

In case a substrate has more than one double bond, selective reduction is possible. Some examples are:

Limonene   p-meth 1(2) ene

p-cyanostyrene   p-cyanoethylbenzene

$$O_2NCH_2CH=CHCH_3 \xrightarrow{\text{Pd/H}_2} O_2NCH_2CH_2CH_2CH_3$$

1-Nitrobut-2-ene   1-Nitrobutane

A selective reducing agent is di-imide (HN=NH). It is prepared in situ by reduction of tosyl hydrazine in boiling xylene.

$$C_7H_7SO_2NHNH_2 \xrightarrow[\text{reflux}]{\text{xylene}} [NH=NH]$$

p-Toluene Sulfonyl hydrazine   Diimide

The reduction of alkenes using di-imide is stereo selective and takes place by *cis* addition of hydrogen in all cases.

2.6.7.1.2   Reduction of Alkynes

Alkynes on reduction can give either fully saturated hydrocarbons (alkanes) or partially saturated hydrocarbons (alkenes) depending on the catalyst employed and reaction conditions.

$$H_3CC\equiv CCH_3 \xrightarrow{\text{Pt/H}_2} [H_3CCH=CHCH_3] \xrightarrow{\text{Pt/H}_2} H_3CCH_2CH_2CH_3$$

2-Butyne   *n*-Butane

Partial reduction of alkynes to alkenes can be affected by reduction with **Lindlar's catalyst** (Pd/CaCO$_3$/lead acetate)

$$CH_3C \equiv CCH_3 \xrightarrow[H_2]{Pd/CaCO_3/lead\ acetate} CH_3CH = CHCH_3$$

2–Butyne   2–Butene

Internal alkynes on reduction can give Z (*cis*) or E (*trans*) alkene depending on the catalyst used. Thus,

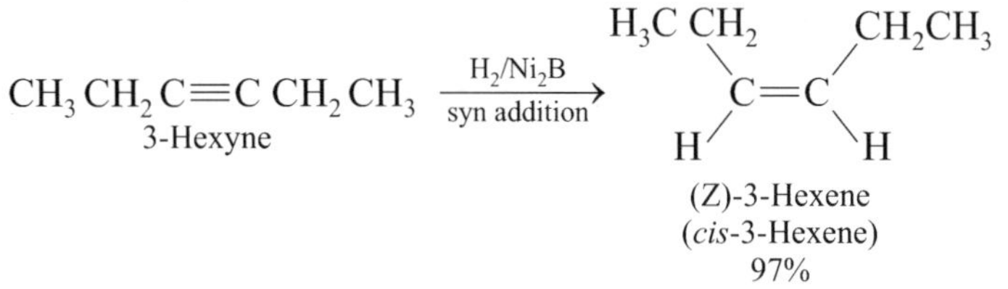

$$CH_3 CH_2 C \equiv C\ CH_2 CH_3 \xrightarrow[syn\ addition]{H_2/Ni_2B}$$

3-Hexyne

H$_3$C CH$_2$ CH$_2$CH$_3$
        \   /
         C = C
        /   \
       H   H

(Z)-3-Hexene
(*cis*-3-Hexene)
97%

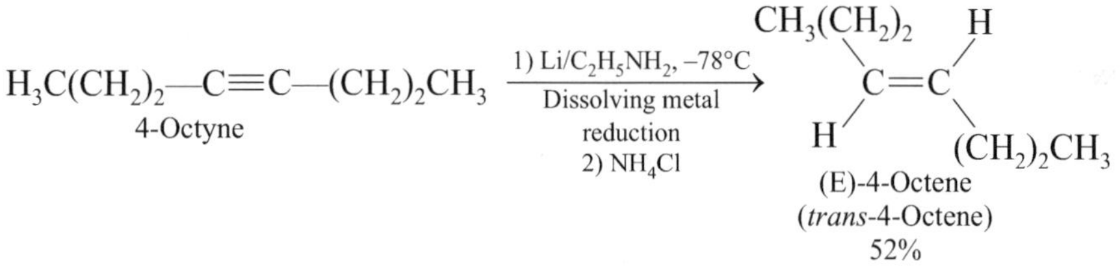

$$H_3C(CH_2)_2 - C \equiv C - (CH_2)_2CH_3 \xrightarrow[\substack{Dissolving\ metal \\ reduction \\ 2)\ NH_4Cl}]{1)\ Li/C_2H_5NH_2,\ -78°C}$$

4-Octyne

CH$_3$(CH$_2$)$_2$ H
         \   /
          C = C
         /   \
        H   (CH$_2$)$_2$CH$_3$

(E)-4-Octene
(*trans*-4-Octene)
52%

2.6.7.1.3   Reduction of Aromatic Hydrocarbons

Aromatic hydrocarbons are difficult to reduce. These can however be reduced by Ni/H$_2$ or PtO$_2$/H$_2$ in acetic acid via the intermediate formation of cyclohexadiene and cyclohexene.

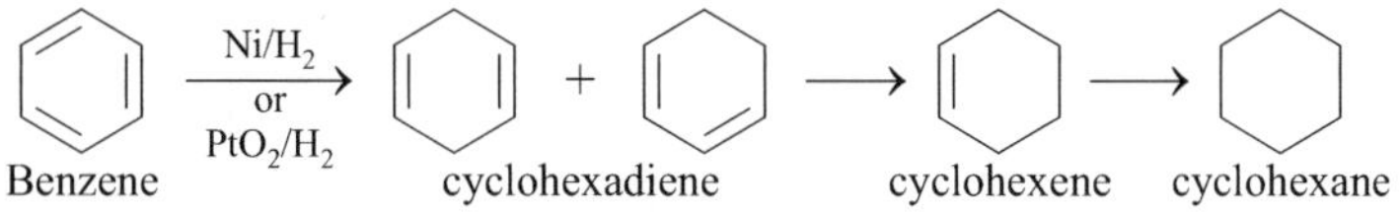

Benzene   cyclohexadiene   cyclohexene   cyclohexane

Benzene can be reduced to 1, 4-hexadiene by sodium (or lithium or potassium) and liquid ammonia and alcohol (**Birch reduction**)

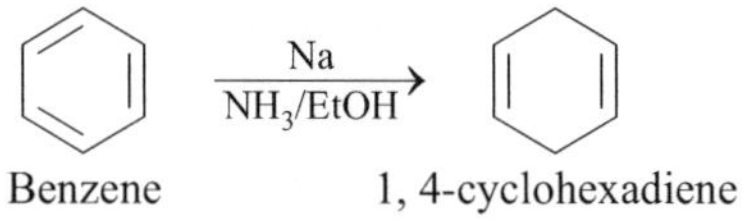

Benzene   1, 4-cyclohexadiene

This is the case of dissolving metal reduction. The mechanism of the reaction is as given below:

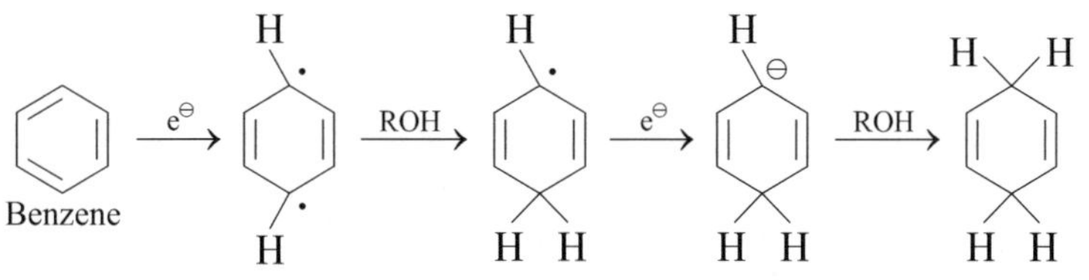

Benzene

The course of Birch reduction is influenced by the subsequent groups in benzene ring. Thus, anisole on Birch reduction gives 1-methoxy-1, 4-cyclohexadiene, which is used for the synthesis of 2-cyclohexenones.

Anisole	1-methoxy-1, 4-cyclohexadiene (85%)	2-cyclohexenone

Condensed aromatic compounds such as naphthalene, anthracene and phenanthacene can be reduced catalytically or with Na/alcohol to give products depending on the reagent used.

Anthracene can be reduced catalytically or with sodium/alcohol.

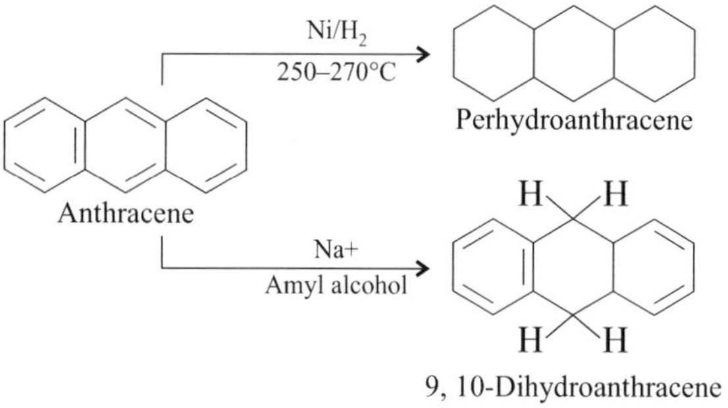

Anthracene

Ni/H$_2$
250–270°C

Perhydroanthracene

Na+
Amyl alcohol

H   H

H   H

9, 10-Dihydroanthracene

Phenanthrene on reduction with sodium/alcohol gives 9,10-dihydrophenanthrene.

Phenanthrene

Na/EtOH
Δ

9, 10-Dihydrophenanthrene

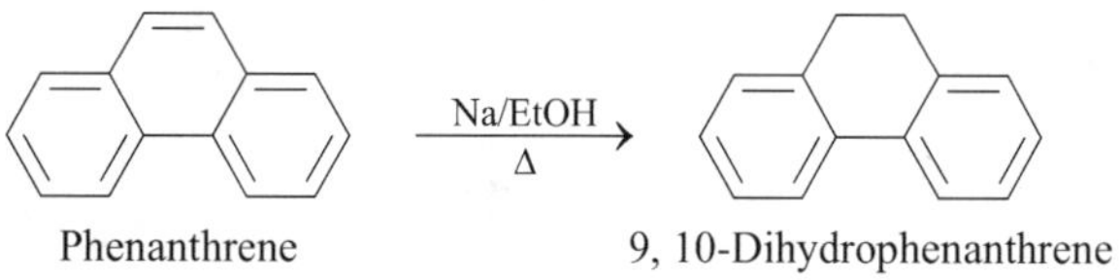

2.6.7.2 Reduction of Alcohols and Phenols

Alcohols are difficult to reduce. However, catalytic hydrogenation of alcohols (1°, 2° or 3°) using molydenum or tungsten sulfides at 300–350°C and 75-125 atm. pr give the corresponding hydrocarbons.

Alternatively, alcohols can be reduced to the corresponding hydrocarbons by treatment with N, N-dicyclohexylcarbodiimide (DCC) followed by catalytic hydrogenation of the formed o-alkoxy–N, N′-dialkylcycloisourcas over palladium at 50-80°C at 1-55 atm. pr.

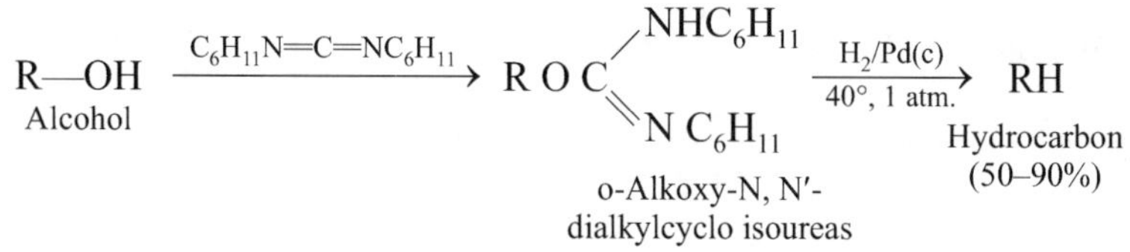

R—OH
Alcohol
$\xrightarrow{C_6H_{11}N=C=NC_6H_{11}}$
$RO\,C\overset{\displaystyle NHC_6H_{11}}{\underset{\displaystyle N\,C_6H_{11}}{\diagdown}}$
$\xrightarrow[40°, \text{ 1 atm.}]{H_2/Pd(c)}$
RH
Hydrocarbon
(50–90%)

o-Alkoxy-N, N′-
dialkylcyclo isoureas

The above method can also be used for the reduction of phenols. Alternatively, phenols can be reduced by converting them into the corresponding tosyl ether and subsequent reduction with Raney Ni/H$_2$. This is known as **hydrogenolysis**.

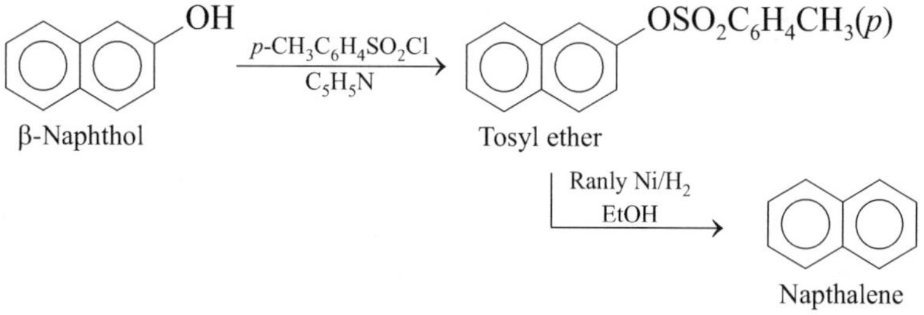

β-Naphthol

$\xrightarrow[C_5H_5N]{p\text{-}CH_3C_6H_4SO_2Cl}$

OSO$_2$C$_6$H$_4$CH$_3$(p)

Tosyl ether

Ranly Ni/H$_2$
EtOH

Napthalene

2.6.7.3 Reduction of Aldehydes and Ketones

2.6.7.3.1 Reduction to Alcohols

Both aldehydes and ketones can be reduced to primary and secondary alcohols respectively by catalytic hydrogenation (Pd/H_2), Sodium-alcohol (Bouveault-Blanc reduction), Sodium borohydride or lithium aluminium hydride.

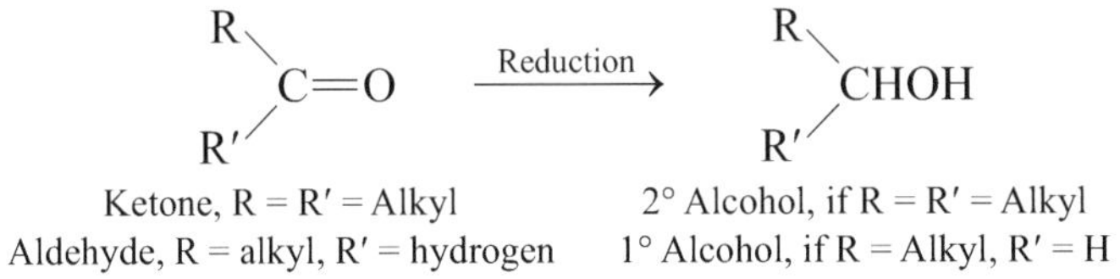

$$R\!\!\diagdown \atop R'\!\!\diagup C=O \quad \xrightarrow{\text{Reduction}} \quad R\!\!\diagdown \atop R'\!\!\diagup CHOH$$

Ketone, R = R′ = Alkyl   2° Alcohol, if R = R′ = Alkyl
Aldehyde, R = alkyl, R′ = hydrogen   1° Alcohol, if R = Alkyl, R′ = H

The mechanism of reduction involving sodium-alcohol (Bouveault-Blanc reduction), $NaBH_4$ reduction and Li AlH_4 reduction and given below:

Mechanism of Bouveault-Blanc Reduction

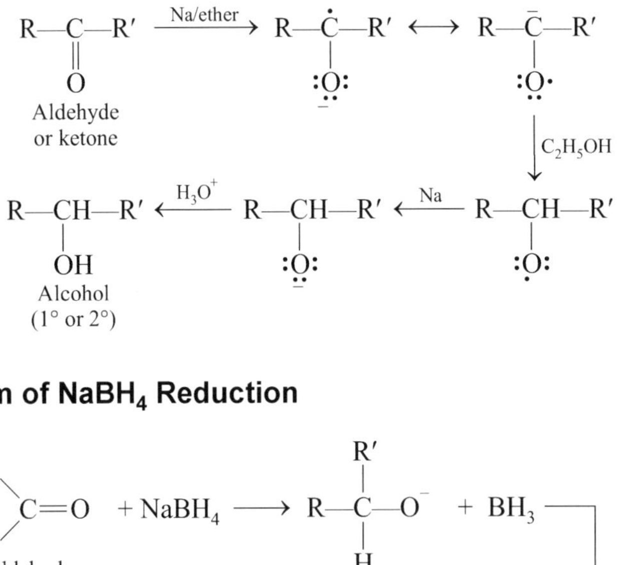

Mechanism of NaBH₄ Reduction

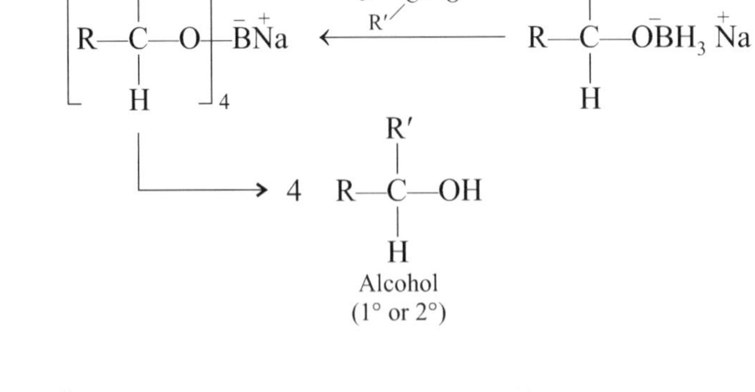

Mechanism of LiAl₄ Reduction

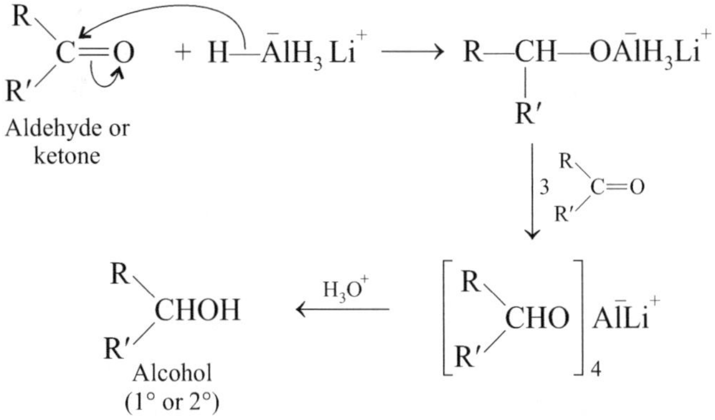

R—CH—OAlH₃Li⁺ formed from aldehyde or ketone + H—ĀlH₃Li⁺

$$\begin{array}{c} R \\ \diagdown \\ C=O \\ \diagup \\ R'' \end{array} + H\text{—}\bar{A}lH_3\,Li^+ \longrightarrow R\text{—}CH\text{—}O\bar{A}lH_3Li^+$$

Aldehyde or ketone

$$\begin{array}{c} R \\ \diagdown \\ CHOH \\ \diagup \\ R'' \end{array} \xleftarrow{H_3O^+} \left[\begin{array}{c} R \\ \diagdown \\ CHO \\ \diagup \\ R'' \end{array}\right]_4 \bar{A}lLi^+$$

Alcohol
(1° or 2°)

Aldehydes and ketones also react with grignard reagent to form an adduct, which on hydrolysis gives the corresponding alcohol.

$$\begin{array}{c} R \\ \diagdown \\ C=O \\ \diagup \\ R'' \end{array} + CH_3MgI \longrightarrow \left[\begin{array}{c} CH_3 \\ | \\ R\text{—}C\text{—}\bar{O}\overset{+}{M}gI \\ | \\ R' \end{array}\right] \longrightarrow \begin{array}{c} CH_3 \\ | \\ R\text{—}C\text{—}OH \\ | \\ R' \end{array}$$

Aldehyde or ketone

Alcohol (2° or 3°)

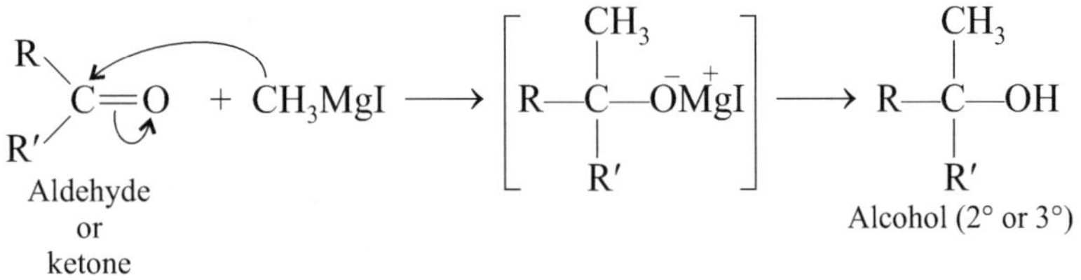

For more details see section 2.6.1.2.5.

Aldehydes and ketones can also be reduced to the corresponding alcohol by **Meerwein-Ponndorf-verley reduction**. The reaction involves heating the aldehyde or ketone with aluminium isopropoxide in presence of isopropyl alcohol

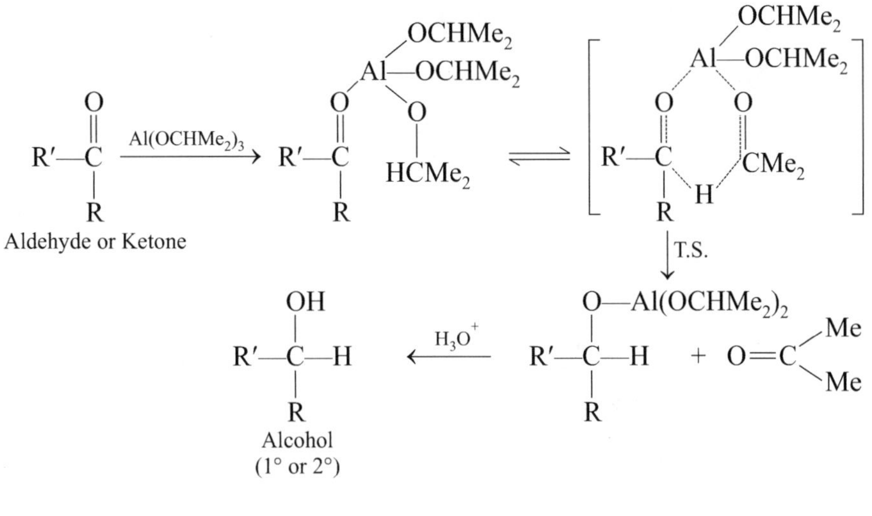

Aldehyde or Ketone

T.S.

Alcohol
(1° or 2°)

2.6.7.3.2 Reduction of Aldehydes and Ketones to Hydrocarbons

Both aldehydes and ketones can be reduced to the hydrocarbons by the following methods:

(*i*) **Clemmensens Reduction:** It involves the treatment of aldehydes or ketones with zinc amalgam and conc. HCl

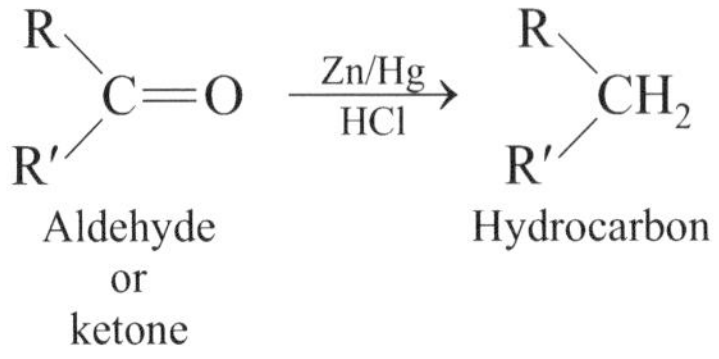

$$\underset{\substack{\text{Aldehyde} \\ \text{or} \\ \text{ketone}}}{\overset{R}{\underset{R''}{\diagup}}C=O} \xrightarrow[\text{HCl}]{\text{Zn/Hg}} \underset{\text{Hydrocarbon}}{\overset{R}{\underset{R''}{\diagup}}CH_2}$$

Mechanism

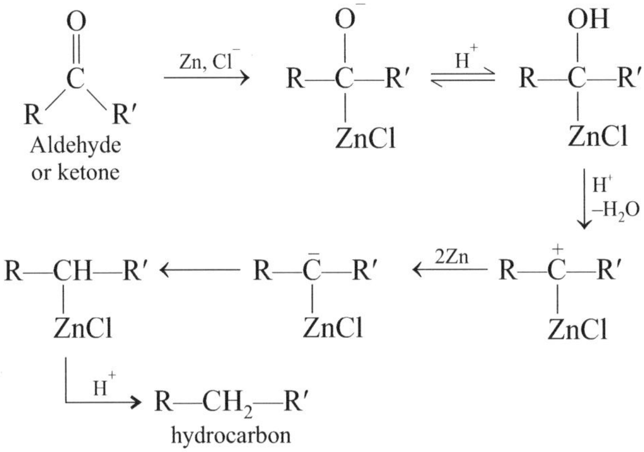

(*ii*) **Wolff-Kishner Reaction:** It involves converting the aldehydes or ketones to the corresponding hydrazone derivative, which on heating with NaOH or KOH give the corresponding hydrocarbon.

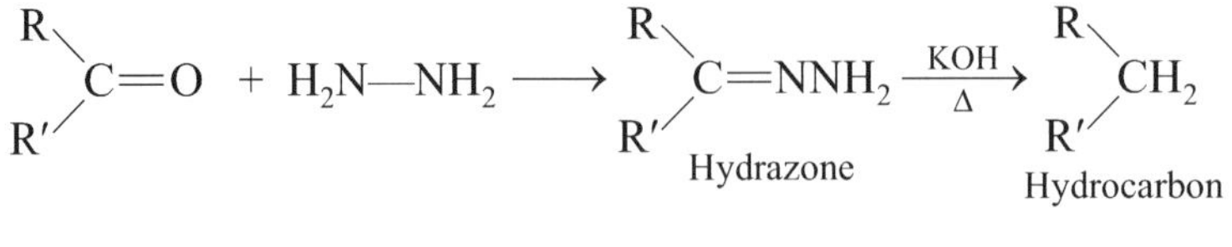

$$\overset{R}{\underset{R''}{\diagup}}C=O \; + \; H_2N-NH_2 \longrightarrow \underset{\text{Hydrazone}}{\overset{R}{\underset{R''}{\diagup}}C=NNH_2} \xrightarrow[\Delta]{\text{KOH}} \underset{\text{Hydrocarbon}}{\overset{R}{\underset{R''}{\diagup}}CH_2}$$

Mechanism

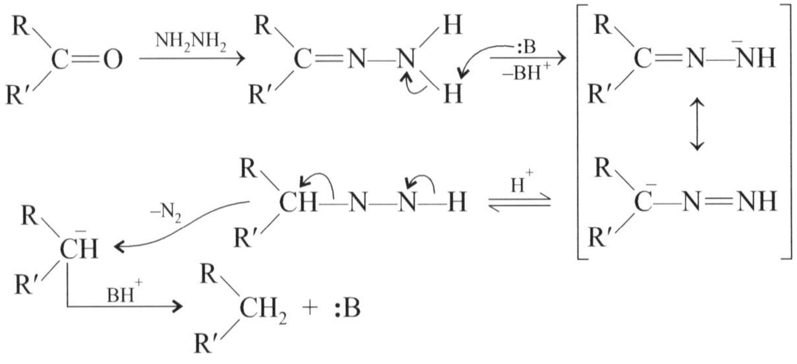

2.6.2.7.4 Reduction of Carboxylic acids and its derivatives

(*i*) **Reduction of carboxylic acids to aldehydes:** Carboxylic acids can be reduced to the corresponding aldehydes by lithium aluminium hydride or aminoalane (prepared in situ by treating LiAlH$_4$ with aluminium chloride)

$$CH_3(CH_2)_nCOOH \xrightarrow[\substack{\text{or} \\ \text{Aminoalane} \\ \text{THF, 0–20°C}}]{LiAlH_4} CH_3(CH_2)_nCHO$$

Hexanoic acid, $n = 4$ Hexanal
Octanoic acid, $n = 6$ Octanal
(60-70%)

Carboxylic acids can also be converted into aldehydes by conversion into acid chlorides followed by reduction with H$_2$/Pd/BaSO$_4$ (**Rosenmund reaction**)

$$RCOOH \xrightarrow{SOCl_2} RCOCl \xrightarrow{H_2/Pd/BaSO_4} RCHO + HCl$$

Reduction of Carboxylic acids to alcohols.

Carboxylic acids can be reduced to the corresponding alcohol by LiAlH$_4$ in ether solution. As the carboxylic acids contains an acidic hydrogen, an additional equivalent of LiAlH$_4$ is required beyond the amount for reduction. The stoichiometric ratio is 4 mol of acid to 3 mol of LiAlH$_4$.

$$4RC\overset{O}{\underset{OH}{\big\langle}} + LiAlH_4 \longrightarrow 4\left[RC\overset{O}{\underset{OH}{\big\langle}}\right]^- \overset{3+}{Al}\,\overset{+}{Li} + 4H_2$$

$$4\left[RC\overset{O}{\underset{O}{\big\langle}}\right]^- \overset{3+}{Al}\,\overset{+}{Li} + 2\,LiAlH_4 \longrightarrow 4\,[RCH_2O]^- \overset{3+}{Al}\,\overset{-}{Li} + 2LiAlO_2$$

$$\Big\downarrow {}_{2H_2O}$$

$$4\,RCH_2OH + LiAlO_2$$

$$4RCOOH + 3LiAl_4 + 2H_2O \longrightarrow 4\,RCH_2OH + 3LiAlO_2 + 4H_2$$

Thus, trimethyl acetic acid and stearic acid are reduced by LiAlH$_4$ to neopentyl alcohol and 1-octadecanol respectively.

$$(CH_3)_3CCOOH \xrightarrow{LiAlH_4/ether} (CH_3)_3CCH_2OH$$

Trimethylacetic acid   Neopentyl alcohol
(92%)

$$CH_3(CH_2)_{16}COOH \xrightarrow{LiAlH_4/ether} CH_3(CH_2)_{16}CH_2OH$$

Stearic acid   1-octadecanol
(Stearyl alcohol)
(91%)

Sebacic acid, a dicarboxylic acid is reduced by LiAlH$_4$ to 1, 10-decanediol in 91% yield

$$HOOC(CH_2)_6COOH \xrightarrow{LiAlH_4/ether} HOH_2C(CH_2)_6CH_2OH$$

Sebacic acid   1,10-Decanediol

Lithium aluminium hydride reduces exclusively the carboxyl group even in an unsaturated acid with an α,β-conjugated double bond. An example is given below:

$$CH_3CH=CHCH=CHCO_2H \xrightarrow{LiAlH_4/ether} CH_3CH=CHCH=CHCH_2OH$$

Sorbic acid
(2,4-Hexadienoic acid)

Sorbic alcohol
(2,4-Hexadien-1-ol)
(92%)

Another example of the reduction of unsaturated carboxylic acid is:

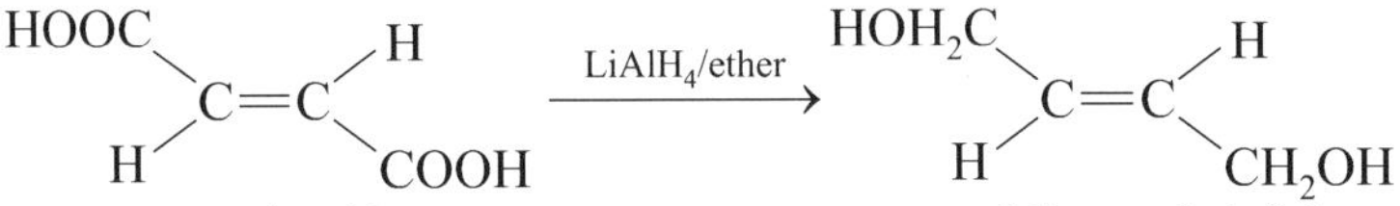

Fumaric acid
(*trans*-1, 2-ethylene
dicarboxylic acid)

trans-2-Butene-1, 4-diol
(78%)

Reduction of Derivatives of Carboxylic Acids

The derivatives of carboxylic acids include acid chloride, anhydride, ester, amides.

As already stated acid chlorides can be reduced to the corresponding aldehyde by Rosenmund reduction (see page 187)

Anhydrides of monocarboxylic acids can be reduced to either lactone or diol depending on the reducing agent. An example is given below:

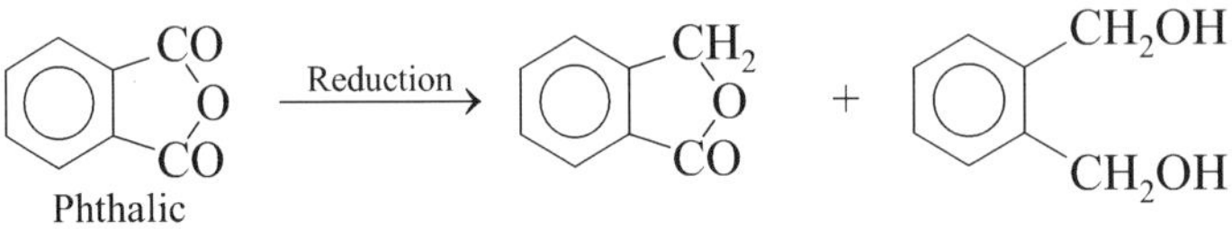

Phthalic
anhydride

$H_2/CuCr_2O_4$, C_6H_6, 260°, 215 atm	82.5%	—
LiAlH$_4$/ether	—	87%
NaAlH$_2$(OCH$_2$CH$_2$OMe)$_2$, C$_6$H$_6$ 80°, 1.5 hr.	—	88.5%

Esters are reduced by sodium/alcohol to the corresponding primary alcohol. The reaction is known as **Bouveault-Blanc reduction**. During the reaction with Na/alcohol the double bonds remain uneffected. An example is given below:

$$CH_3(CH_2)_7CH=CH(CH_2)_7CO_2H \xrightarrow[reflex/1\ hr.]{Na/EtOH} CH_3(CH_2)_7CH=CH(CH_2)_7CH_2OH$$

Ethyl oleate

Oleyl alcohol (49–51%)

H_2, $CuCr_2O_4$, 250°,200 atm
80% yield

Esters can also be reduced to aldehydes by DIBAL-H, Diisobutyalanane or diisobutylaluminium hydride [$AlH(CH_2CHMe_2)_2$]. An example is given below:

The reduction of esters to primary alcohols can be best accomplished by $LiAlH_4$ or $iBu\,AlH_4$ or $LiBH_4$.

1° Alcohol

Amides on reduction with Na/alcohol or $LiAlH_4$ yield 1° amines.

$$RCONH_2 \xrightarrow[\text{or Na/EtOH}]{LiAlH_4} RCH_2NH_2$$

Amide   1° Amine

2.6.2.7.5   Reduction of Cyanides

Aromatic cyanides on reduction with stannous chloride/HCl followed by hydrolysis give the corresponding aldehydes. The reaction is known as **stephen's reaction**

$$C_6H_5C{\equiv}N \xrightarrow[\text{(2) hydrolysis}]{\text{(1) } SnCl_2/HCl} C_6H_5CHO$$

Mechanism

$$RCN \xrightarrow[HCl]{SnCl_2} [R{-}C{\equiv}\overset{+}{N}H]Cl^- \xrightarrow{[H]} R{-}CH{=}\overset{+}{N}H_2Cl^-$$

$$\downarrow H_2O$$

$$RCHO$$

2.6.2.7.6   Reduction of Aromatic Nitro Compounds

Reduction of nitro compounds with H_2/PtO_2 or Raney Ni/H_2 give the corresponding primary amines.

$$R-NO_2 \xrightarrow[EtOH]{H_2/PtO_2 \text{ or Raney Ni}/H_2} RNH_2$$

Nitro compound   i Amine

Both aliphatic and aromatic compounds can be reduced with Fe/HCl or $LiAlH_4$

$$RNO_2 \xrightarrow[\text{or } LiAlH_4/ether]{Fe/HCl} RNH_2$$

Selective reduction of nitro group can be effected with ammonium sulphide.

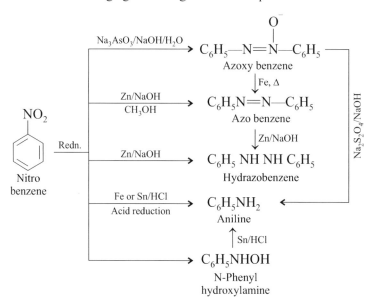

m-Dinitrobenzene   *m*-Nitroaniline

In case of nitrobenzene a typical nitro compound, depending on the reaction conditions and the reducing agent used give different products.

MULTIPLE CHOICE QUESTIONS

1. In an organic molecule each of the two hybridized orbital has 33% *s* and 66% *p* character. The organic compound is
   (*a*) Methane   (*b*) ethene
   (*c*) ethyne   (*d*) can be any of the above

2. A Substituent having a negative inductive effect (– I effect) is
   (*a*) OH^- (*b*) $(CH_3)_3C^-$
   (*c*) $CH_3CH_2^-$ (*d*) CH_3^-

3. A substitutent having a positive inductive effect (+ I effect) is
   (*a*) $C_6H_5^-$ (*b*) NO_2^-
   (*c*) $COOH^-$ (*d*) CH_3^-

4. In alkyl halides, the inductive effect can explain
   (*a*) effect on bond length   (*b*) effect on dipole moment
   (*c*) reactivity of alkyl halides (*d*) all the above

5. Out of the following carboxylic acids, the strongest one is:

   (a) Acetic acid　　　　　　　(b) Chloroacetic acid

   (c) Cyanoacetic acid　　　　　(d) all are equally strong

6. The actual relative basic strength of the following amines is of the order

   (a) 1° amine > 2° amine > 3° amine

   (b) 2° amine > 1° amine > 3° amine

   (c) 3° amine > 2° amine > 1° amine

   (d) all amine are of same basic strength.

7. The group having maximum hyperconjugate release is

   (a) tertiary butyl group

   (b) isopropyl group

   (c) ethyl group

   (d) All are having equal hyperconjugative release

8. Out of the following three carbocations, the most stable is

   (a) allyl carbocation

   (b) benzyl carbocation

   (c) Propyl carbocation, $CH_3CH_2\overset{+}{C}H_2$

   (d) all carbocation are equally stable.

9. The reaction of HBr on neopentyl alcohol [$(CH_3)_3C$—CH_2OH] gives 2-bromo-2-methyl butane [$(CH_3)_2$—$C(Br)$—CH_2—CH_3]. This rearrangement is called

   (a) Wagner–Meerwein rearrangement

   (b) Pinacol–Pinacolone rearrangement

   (c) Beckmann rearrangement

   (d) Any of the above|

10. The stability of the following three carbanions is of the order

    (a) $CH_3\bar{C}H_2 > CH_2{=}\bar{C}H > HC{\equiv}\bar{C}$

    (b) $CH_2{=}\bar{C}H > CH_3\bar{C}H_2 > HC{\equiv}\bar{C}$

    (c) $HC{\equiv}\bar{C} > CH_2{=}\bar{C}H > CH_3{-}\bar{C}H_2$

    (d) all are of equal stability

11. The reaction of R_3C—CHO with CH_3CHO in presence of a base gives

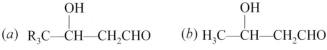

    (a) R_3C—$\overset{\overset{\displaystyle OH}{|}}{C}H$—$CH_2CHO$　　(b) H_3C—$\overset{\overset{\displaystyle OH}{|}}{C}H$—$CH_2CHO$

    (c) Both (a) and (b)　　　　　(d) There is no reaction

12. The stability of the 1°, 2° and 3° free radicals is of the order
    (a) 1° > 2° > 3°   (b) 3° > 2° > 1°
    (c) 2° > 1° > 3°   (d) all are of equal stability

13. Out of the following three free radicals the most stable is
    (a) $Ph_2\dot{C}H$ (b) $Ph\dot{C}H_2$
    (c) $Ph_3\dot{C}$ (d) all are of equal stability

14. $Ph_2\dot{N}$ is an example of
    (a) bridgehead free radical   (b) hetero radical
    (c) Phenoxy radical   (d) Thiyl radical

15. The following reaction

$$Ar-\overset{\overset{O}{\|}}{C}-O-O-\overset{\overset{O}{\|}}{C}-Ar + Cu^+ \longrightarrow Ar-\overset{\overset{O}{\|}}{C}-\dot{O} + ArCO_2^- + Cu^{2+}$$

    is an example of
    (a) Thermolysis reaction   (b) Photolysis reaction
    (c) Redox reaction   (d) can be any of the above

16. The product obtained by heating ethyl free radical $(CH_3\dot{C}H_2)$ is
    (a) $CH_2{=}CH_2$ (b) $CH_3{-}CH_3$
    (c) a mixture of (a) and (b)   (d) none of the above

17. The electrolysis of sodium or potassium salts a carboxylic acid to form an alkane proceeds via the formation of

    (a) Alkoxy radical, $R-\overset{\overset{O}{\|}}{C}-\dot{O}$ (b) Alkyl radical $\dot{R}$
    (c) Both (a) and (b)   (d) only alkoxy radical

18. The reaction of excess of methane and chlorine in presence of sun light gives
    (a) CH_3Cl (b) CCl_4
    (c) CH_2Cl_2 (d) a mixture of (a), (b) and (c)

19. The reaction of toluene with chlorine in presence of sun light gives benzyl chloride. The reaction proceeds via the formation of
    (a) Benzyl radical   (b) Hexadecyl radical
    (c) both (a) and (b)   (d) none of the above radicals

20. The carbon atom in a carbene is an electron deficient species since it has ____ electrons
    (a) 2   (b) 3
    (c) 5   (d) 6

21. The reaction of trans-2-butene with CH_2N_2 in presence of sunlight gives

(a)

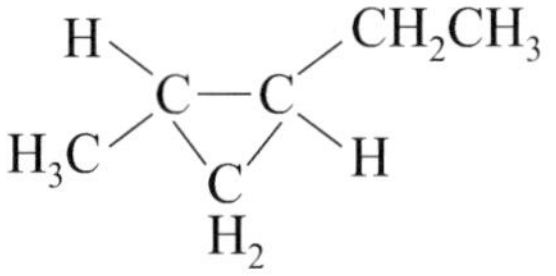

$$\underset{H_3C}{\overset{H}{\diagdown}}C-C\underset{\underset{\substack{C\\H_2}}{H}}{\overset{CH_2CH_3}{\diagup}}$$

(b)

$$\underset{H_3C}{\overset{H}{\diagdown}}C=C\underset{H}{\overset{CH_2CH_3}{\diagup}}$$

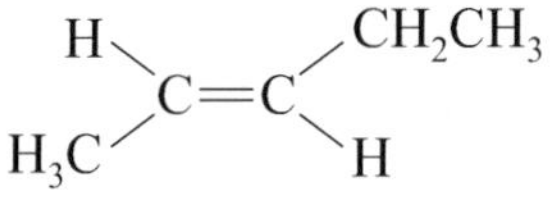

(c) a mixture of (a) and (b) (d) there is no reaction

22. Nitriles are obtained by the reaction of dichlorocarbene with

(a) $RCO\,NH_2$ (b) $RCS\,NH_2$

(c) $RCH=NOH$ (d) $R-\underset{\underset{NH}{\|}}{C}-NH_2$

(e) any of the above

23. The product obtained in the following reaction is

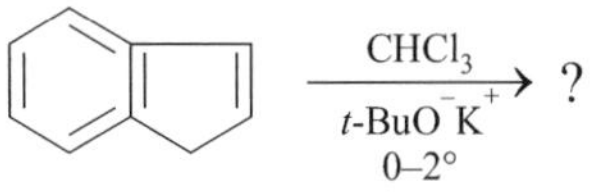

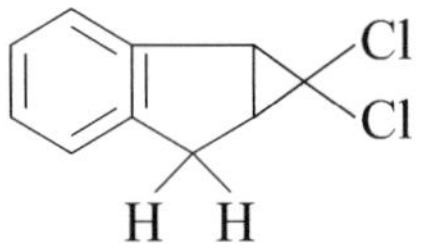

$$\xrightarrow[\substack{t\text{-BuO}^-\text{K}^+\\0\text{–}2°}]{CHCl_3} ?$$

(a) [structure with Cl, Cl and H H]

(b) [naphthalene structure with Cl]

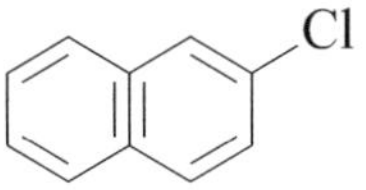

(c) a mixture of (a) and (b) (d) There is no reaction

24. The major product obtained in Reimer-Tiemann reaction is

(a) o-Hydroxy benzaldehyde

(b) p-hydroxy benzaldehyde

(c) a mixture of 1 : 1 (a) and (b)

(d) There is no reaction

25. The reaction of p-chlorotoluene with KNH_2 gives

(a) p-Toluidine (b) m-Toluidine

(c) a mixture of (a) and (b) (d) there is no reaction

26. m-Aminoanisole can be obtained by the reaction of _____ with $\overset{+}{K}\overset{-}{NH_2}$.

(a) o-Bromoanisole (b) m-Bromoanisole

(c) either (a) or (b) (d) there is no reaction

27. Which of the following is a neutral nucleophile

(a) BF_3 (b) NH_3

(c) H_3O^+ (d) $\diagdown C=O$

28. The reaction of ethylene with bromine to give dibromoethane, $BrCH_2-CH_2Br$ is an example of

(a) electrophilic addition reaction

(b) Free radical addition reaction

(c) Nucleophilic addition reaction

(d) none of the above

29. The reaction of propene with HBr gives

   (*a*) 1-Bromopropane

   (*b*) 2-Bromopropane

   (*c*) A mixture (1 : 1) of (*a*) and (*b*)

   (*d*) there is not reaction

30. Addition of HBr to propenenitrile gives

   (*a*) 2-Bromopropane nitrile

   (*b*) 1-Bromopropane nitrile

   (*c*) A 1 : 1 mixture of (*a*) and (*b*)

   (*d*) No reaction takes place.

31. The reaction of 2-pentene with HBr gives

   (*a*) $CH_3CH_2CH\!-\!\overset{\displaystyle Br}{\underset{\displaystyle H}{|}}CHCH_3$
   (*b*) $CH_3CH_2\overset{\displaystyle Br}{\underset{\displaystyle H}{|}}CH\!-\!CHCH_3$

   (*c*) a mixture of (*a*) and (*b*)   (*d*) there is no reaction

32. *tran* hydroxylation of alkene can be effected by

   (*a*) Reaction with osmium tetraoxide in presence of an oxidizing agent

   (*b*) oxidation of alkene with dil. $KMnO_4$ solution

   (*c*) Acid catalysed ring opening of epoxide ring

   (*d*) Oxymercuration-demercuration of alkene

33. The halogenation of alkenes gives mainly

   (*a*) trans additions product

   (*b*) *cis*-addition product

   (*c*) a 1 : 1 mixture of (*a*) and (*b*)

   (*d*) No reaction takes place

34. The addition of bromine on butadiene gives

   (*a*) 3, 4-Dibromo-1-butene   (*b*) 1, 4-Dibromo-2-butene

   (*c*) a mixture of (*a*) and (*b*)   (*d*) no reaction takes place

35. 1-Bromopropane nitrile is obtained from propene by

   (*a*) addition of HBr

   (*b*) addition of HBr in presence of peroxide

   (*c*) any of the methods described in (*a*) and (*b*)

   (*d*) cannot be obtained either by (*a*) or (*b*)

36. In Bimolecular Elemination reactions (E_2), the rate of the reaction increase in case of alkyl halides in the order

   (*a*) $3° > 2° > 1°$ (*b*) $1° > 2° > 3°$

   (*c*) $2° > 1° > 3°$ (*d*) All alkyl halides react by the same rate

37. The order of reactivity of RX in an bimolecular reaction is of the order
    (a) R—F > R—Cl > R—Br > RI
    (b) RCl > RBr > RF > RI
    (c) RBr > R—Cl > R—I > RF
    (d) R—I > R—Br > R—Cl > R—F

38. The reaction CH_3—CH—CH_2—CHO $\xrightarrow{\text{base}}$ CH_3CH=CH—CHO

    takes place by
    $\overset{\displaystyle |}{\underset{\displaystyle OH}{}}$
    (a) E_2 mechanism   (b) E_1 mechanism
    (c) ElCB mechanism   (d) any of the above mechanism

39. The reactivity in aromatic electrophilic substitution depends on the presence of an
    (a) activating group
    (b) deactivating group
    (c) both (a) and (b)
    (d) does not depends on any of the above groups

40. The substitution of bromine by methoxy group in (L)-2-bromopropionate involving retention in configuration can be explained by
    (a) S_N2 mechanism
    (b) S_N1 mechanism
    (c) participation of a neighbouring group
    (d) none of the above

41. The order of reactivity in nucleophilic aromatic-substitution in o-nitrotoluene (I) 2, 4-dinitrotolune (II) and 2, 4, 5-trinitrotoluene (III) is of the order
    (a) I > II > III   (b) III > II > I
    (c) II > I > III   (d) III > I > II

42. Pinacol-Rearrangement is a type of
    (a) carbon-carbon rearrangement
    (b) carbon-nitrogen rearrangement
    (c) carbon-oxygen rearrangement
    (d) none of the above

43. The conversion of 3-Hydroxy-1, 5-hexadiene into 1-hexanal on heating is an example of
    (a) Isomerisation   (b) cope rearrangement
    (c) Electrocyclic reaction   (d) photochemical reaction

44. The photochemical reaction of butadiene in presence of acetophenone give the major product.

(a)

(b)

(c)

(d) all are formed in equal amounts

45. Primary alcoholic group in complex compound containing other oxidisiable group and asymmetric centre can be oxidised to the corresponding aldehydes using

(a) $KMnO_4$

(b) H_2O_2/alkate

(c) Cr_2O_3/H^+

(d) Tetra-n-propyl ammonium perruthenate

Answers

1. (b)	2. (a)	3. (d)	4. (d)	5. (c)
6. (b)	7. (a)	8. (b)	9. (a)	10. (c)
11. (c)	12. (b)	13. (c)	14. (b)	15. (c)
16. (c)	17. (c)	18. (a)	19. (a)	20. (d)
21. (c)	22. (e)	23. (b)	24. (a)	25. (c)
26. (c)	27. (b)	28. (a)	29. (b)	30. (b)
31. (c)	32. (c, d)	33. (a)	34. (c)	35 (c)
36. (a)	37. (d)	38. (c)	39. (c)	40. (c)
41. (b)	42. (a)	43. (b)	44. (b)	45. (d)

FILL IN THE BLANKS

1. Out of 1°, 2° and 3° amines, the least basic is _____.

2. Out of o-, m- and p- nitrobenzoic acids, the most acidic is _____.

3. Out of aniline, ammonia and cyclohexylamine, the most weaker is _____.

4. The free radical which is most stable is _____ free radical.

5. Friedel crafts reaction of benene with n-propyl chloride in presence of anhyd. $AlCl_3$ gives _____.

6. The reaction of acetophenone with perbenzoic acid in $CHCl_3$ at 25°C gives _____.

7. The stability of the benzyl carbocation (I), vinyl carbocation (II) and cyclopentyl carbocation (III) is of the order _____.

8. The reaction of triphenyl methyl chloride with sodium followed by reaction with CO_2 gives _____.

9. The reaction of β-phenyl ethyl bromide with a base gives _____.

10. Diphenyl disulphide on heating gives _____.

11. The reaction of the aldehyde, $Ph_2MeC\ CH_2\ CHO$ with the radical $Me_3\overset{.}{C}O$ gives _____.

12. Aniline on diazotisation followed by reaction with CuBr at 100° yields _____.

13. The above reaction in 12 proceeds via _____ free radical.

14. Alkenes on epoxidation followed by photolytic decomposition of the formed epoxide gives _____.

15. The reaction of primary amines with dichloro carbene yield _____.

16. The reaction of indole with dichlorocarbane gives _____.

17. In Hofmann rearrangement, the intermediate reaction intermediate is a _____.

18. Carbonyl group is an example of _____ nucleophile.

19. Organic reaction which take place in one step are called _____ reactions.

20. The reaction of propene with water containing dilute H_2SO_4 gives _____.

21. The halogenation of alkyne with bromine gives mainly the _____ addition product.

22. The reaction of 1, 4-pentadiene with one mole of bromine gives _____.

23. In Diels-Alder reaction, the dienes can react only when it adops the _____ form.

24. In an Bimolecular elimination reaction, the 1, 2-elimination in case of unsymmetrical substrate is determined by _____.

25. In case of quaternary ammonium salts in which there are two possibility of the product is predicted by _____.

26. The carbon atom adjacent to a double bond is called _____.

27. The order of reactivity of the heterocycles pyrrole, Furan and thiophene is of the order _____.

28. The majour product obtained by treatment of R (–)-2-branooctane with a base is _____.

29. Beckmann reaction is an example of _____ rearrangement.

30. 1, 5-Hexadiene on heating gives a degenerate 1, 5-hexadiene. This is an example of _____.

31. 1, 1, 1-Triphenyl-2-branoethane on treatment with lithium metal in THF followed by acidification gives _____.

32. Photochemical cyclisation may proceed by _____ fashion.

33. Conversion of alpha diazo ketone into ketene _____ is an example of _____.

34. Oxidation of 2-methyl-2-butene with KMnO$_4$, on heating followed by acidification gives _____.

35. Treatment of an alkene with an peracid followed by hydrolysis gives _____ diol.

36. Hydroboration of 1-butyne with disiamyl borane followed by oxidation with alkaline H$_2$O$_2$ gives _____.

37. Toluene on treatment with CrO$_2$Cl$_2$ followed by acidification gives _____.

Answers

1. 3° Amines

2. *o*-Nitrobenzoic acid

3. aniline

4. tertiary

5. isopropyl benzene

6. Phenyl acetate

7. I > II > III

8. Ph$_2$—C—CH$_2$Ph
$\quad\quad\quad$ |
$\quad\quad\quad$ COONa

9. Ph CH=CH$_2$

10. PhS˙

11. PhMeC—CH$_2$
$\quad\quad\quad\quad$ |
$\quad\quad\quad\quad$ Ph

12. Bromobenzene

13. Phenyl

14. carbene

15. isonitriles

16. 3-Chloroquinoline

17. nitrene

18. neutral

19. concerted

20. isopropyl alcohol

21. trans

22. 4, 5-dibromo-1-pentene

23. S-*cis*

24. Saytzeff rule

25. Hofmann rule

26. allylic carbon atom

27. pyrrole > furan > thiophene

28. (S) (+) -2-octanol

29. intermolecular

30. Cope rearrangement

31. 1, 1, 2-triphenylethane

32. intramolecular

33. Wolff rearrangement

34. Ethyl methyl ketone

35. trans

36. butanal

37. benzaldehyde

SHORT ANSWER QUESTIONS

1. Explain the term hybridization.

2. What do you understand by electromeric effect.

3. Compare the inductive and mesomeric effects.

4. What is hyperconjugation.

5. Explain the difference between tautomerism and resonance.

6. Homolytic bond fission can take place by single electron shift. Can it also take place by fission of two bonds. It yes, explain.

7. Explain the terms carbocations, carbanions and free radicals.

8. Friedel-crafts reaction of benzene with *n*-propyl chloride gives isopropyl benzene and not the expected *n*-propyl benzene. Explain.

9. In Baeyer-villager oxidation involving the oxidation of acetophenone $(C_6H_5COCH_3)$ to phenyl acetate $(C_6H_5\overset{\overset{O}{\|}}{C}—OCH_3)$, the carbonyl oxygen of the ketone becomes the carbonyl oxygen of the ester. Explain.

10. The following Beckmann rearrangement

$$C_6H_5\overset{\overset{}{\underset{\underset{NOH}{\|}}{C}}}{—}CH_3 \xrightarrow{\ PCl_5\ } ?$$

gives $CH_3CO\,NH\,C_6H_5$ and not $C_6H_5\,CO\,NH\,CH_3$. Explain.

11. Explain the term 'Crossed Aldol condensation'. Give one example.

12. What are bridgehead free radicals. Give one example.

13. How are phenoxy radicals obtained.

14. How benzyloxyfree radical $(C_6H_5\overset{\overset{O}{\|}}{C}—O^{\cdot})$ is generated.

15. What is Fenton's regent.

16. How following conversions brought about
    (*i*) $CH_3CH_2CH_2Br \longrightarrow CH_3(CH_2)_4CH_3$
    (*ii*) $RCOO\,Ag \longrightarrow RBr$

17. Give mechanism of Sandmer reaction and Gomberg reaction.

18. How is silver salt of a carboxylic acid $(R\overset{\overset{O}{\|}}{C}—OAg)$ converted into the corresponding alkyl bromide (RBr).

19. How is propene converted into secondary propyl bromide and *n*-propyl bromide.

20. How is carbene generated.

21. What do you understand by phase transfer catalyst give an application of PTC in organic synthesis.

22. How Ketenes are generated.

23. What are Benzynes. How are they generated.

24. What product is expected to be formed by the reaction of benzyne with furan and primary amines.

25. What are nucleophiles. How are they classified.

26. Explain the formation of 2-bromopropane by the reaction of propene with HBr.

27. The reaction of bromoethene with HBr gives 1, 1-dibromo ethane Explain.

28. What product is obtained by the addition of HBr to NC.C CH=CH$_2$. Explain your answer.

29. The halogenation of alkenes takes place only with Cl$_2$ or bromine and not with iodine and Fluoriene. Explain why it is so.

30. Bromination of butadiene with bromine gives a mixture of 1, 2- and 1, 4-dibromo adducts, give the mechanism involved.

31. How is *n*-propyl bromide obtained by the reaction of propene with HBr, give the mechanism involved.

32. Propene reacts with HBr in presence of peroxide to give anti-Markovnikov product. HCl and HI do not react. Explain why?

33. Explain the reduction of alkene to alkane by hdyroboration-oxidation procedure.

34. Give the mechanism involved in the formation of ozonide by the reaction of an alkene with ozone.

35. How are trans diols obtained from alkenes.

36. How is acrylaldehyde (CH$_2$=CHCHO) can be converted into the corresponding diol (CH$_2$—CH—CHO).
    $\underset{OH}{|}$ $\underset{OH}{|}$

37. Write a note on Witting reaction.

38. How is benzyl alcohol best obtained from benzaldehyde.

39. Discuss the sterochemistry of E$_2$ eliminations.

40. Explain the formation of an α-β-unsaturated aldehyde by the treatment of a β-hydroxyl carbonyl compound with base.

41. Explain how allytic and vinylic substitution takes place.

42. Explain the term 'ortho effect'.

43. Nitration of pyridine with KNO_3/H_2SO_4 at 37°C gives 3-nitropyridine as the major product. How will you obtain 4-nitropyridine.

44. How nucleophilic substitution in alcohols can be brought about.

45. Explain S_N Ar mechanism.

46. Write a note on nucleophilic heteroaromatic substitution.

47. How nucleophilic substitution can be brought about in case of 3-halopyridines.

48. Write a note on Benzil-Benzilic acid rearrangement.

49. What are photochemical reactions. Give an example.

50. Write a note on Barton reaction.

51. What are oxetanes. How are they obtained.

52. Define oxidation and reduction.

53. How is the change in oxidation state of the carbon atom (in the functional group) is helpful to infer whether the reaction is oxidation or reduction reaction.

54. How methane can be converted into methyl alcohol and formaldehyde.

55. Write a note on Etard's reaction.

56. How is toluene converted into cis-2, 3-dihydrocyclohexa-4, 6-dione.

57. How is Des-Martin reagent prepared. Give its uses.

58. Give mechanism of Dakins oxidation.

59. How is dimethyl dioxirane prepared. Give its uses.

60. How is diimide propared. Give it uses.

61. Write a note on Birch reduction.

62. How is naphthalene converted into cis-decalain, trans-decalin and 1, 2-Dialin.

63. Give the mechanism involved in the reduction of an alcohol into the corresponding hydrocarbon using N,N'-dicyclohexylcarbodiimide.

64. Give the mechanism of Baeyer-Villager oxidation.

Index

© The Author(s) 2023
V. K. Ahluwalia, *Organic Reactions and Their Mechanisms*,
https://doi.org/10.1007/978-3-031-15695-3

Printed in the United States
by Baker & Taylor Publisher Services